Green Revolutions:
Adapting, Abundance, Criticisms and Technology

Mark Roberts-Seymour, PEng.

Published by Secular Orders Press

1018 – 4900 – 20th Street
Vernon, BC, Canada V1T 9W3

Library of Congress Cataloguing:
Mark E. Roberts-Seymour, OFS (1948-)
Green Revolutions: Adapting, Abundance, Criticisms and Technology

1. Sociology, 2. Technology of Farming, 3. Title

ISBN 9781717983251
Copyright © 2018 by Mark Roberts-Seymour
**All Rights Reserved
Second Edition**

Green Revolutions

Contents

Green Revolutions: 1
Chapter 1 Introduction 5
Chapter 2 World Population Growth and Malthusian Arguments 9
    Malthus 10
    Neo Malthusians 17
    The Precariat 19
Chapter 3 Longevity and Environmental Mitigants ... 21
Chapter 4 The Green Revolutions 25
    The 1st Green Revolution 25
    Effects 27
    Criticisms 28
    The 2nd Green Revolution 29
    Mono-Culture 30
    Bio-Diversity 33
    Fossil Fuels and Pesticides 36
    Indian Experience 38
Chapter 5 Abundance Theorism 41
    Solar Power 41
    Thorium 42
Non-Traditional Farming 48
    3-D Printing 50
Chapter 6 Farming and 3D Printing 52
    The 3-D Farm-Bot 54
    Costing for Farms 57

- Construction 3-D Printing..................................63
- Climate Change and Farm-Abundance 75
- Chapter 7 Protein Projections..................................78
- Fish Farming...91
- Chapter 8 Genetically Modified Organisms..............97
  - Farming.and Poverty..108
  - North American Experience..............................110
  - Feeding a Hungry Planet 113
  - The Uganda Experience 117
- Chapter 9 GMO Growth..120
  - Bio-remediation..130
  - Post-Food Crop Applications 134
- Chapter 10 Rates of GMO Introduction 138
- Concluding Remarks 149
- About the Author...150

Chapter 1 Introduction

This book addresses four divisible Green Revolutions, each of which is still 'advancing' today. As Agriculture 'practitioners' we must advance. Notably, the pressure is on from climate deterioration.

Professor John Beddington (b.1945) acting as the Chief Scientist of Britain, has warned that growing populations, falling energy reserves and food shortages would create a "perfect storm" of shortages of food, water, and energy by 2030. He has said that there is already enough CO_2 in the atmosphere for there to be more floods and droughts over the next 25 years. Beddington maintains there is a "need for urgency" in tackling growing shortages and climate disturbances.

"The [current] variation we are seeing in temperature or rainfall is double the rate of the average. That suggests that we are going to have more droughts, we are going to have more floods, we are going to have more sea surges and we are going to have more storms".

"These are the sort of changes that are going to affect us in quite a short timescale," he warned. Beddington's comments come at a time when "climate sceptics" have been challenging claims by scientists that the release of CO_2 into the atmosphere is increasing global temperatures. Other critics have argued that even if the burning of fossil fuels is changing the planet's climate, the reduction of CO_2 levels by the world's emerging nations is unrealistic, impractical and undesirable.

Prof Beddington's blunt response is: "The evidence that climate change is happening is completely unequivocal."

But the issue, he says, has been clouded by the fact that the planet's climate system operates slowly to changes and so there are long delays in CO_2 level rises in the atmosphere resulting in changes to weather patterns "So the next 20 or 30 years are going to be determined by what's up there now." Governments have agreed to try to keep the rise in average global temperatures to below 2C. Given the slow progress in attempts to curb CO_2 emissions at successive climate change talks, many experts believe that target to be unrealistic.

John Beddington made his comments in an interview with BBC News to coincide with the end of his tenure as the government's chief scientific adviser. He is among those credited with having limited the cuts to the UK's research budget when the Chancellor, George Osborne, was making his first Autumn Statement in 2010. However, he could not prevent a deep cut of £1.6bn in spending for research equipment, buildings and infrastructure. Beddington has gradually won back that money in a series of announcements of new money for science by the Treasury. But these have been targeted at specific areas of science, such as graphene research, nanotechnology and space, thought to be important to the growth of UK economy.

This has won support for science from the Treasury. But because the funding is directed to these specific areas, there is less money overall to fund staff and cutting-edge research outside those deemed by senior members of the research community as strategically important.

According to a 2009 report by the United Nations Food and Agriculture Organisation (FAO), the world will have to produce 70% more food by 2050 to feed a projected increase of 2.3 billion people. Running parallel, the automation of jobs, beginning with the manufacturing industry, will have major implications for global employment, especially given the pace at which the technology is advancing. With a projection up to 47 percent of jobs in the United States alone at risk of replacement by automation/robotics within twenty years (After Oxford University), extensive job loss is a reality that the world must prepare to face.

Chapter 2 World Population Growth and Malthusian Arguments

In examining the predictable future of mankind, the world population is critical. As of August 2016, the population was estimated at 7.4 billion. Notwithstanding the trend to longer life expectancy, predictions indicating a slowing of growth from 1968 onward.

World population has experienced continuous growth since 1350. The highest population growth rates – global population increases above 1.8% per year – occurred between 1955-1975 peaking to 2.06% between 1965-1970.

The growth rate declined to 1.18% between 2010-2015 and is projected to decline to 0.13% by the year 2100. Total annual births were highest in the late 1980s at about 139 million, and are now expected to remain essentially constant at their 2011 level of 135 million, while deaths number 56 million per year and are expected to increase to 80 million per year by 2040. Selecting 2040 as a benchmark; the net increase is expected to be 55 million persons per annum. That is growth rate of approximately 0.65%, only 30% of the 1970 rate. These rates assume no catastrophic intervention to lower rates.

The median age of the world's population was estimated to be 30.1 years in 2016, with the male median age estimated to be 29.4 years and female, 30.9 years. The world life expectancy at 2015 was 71.4 years.

A 2014 estimate forecasts a 2040 approaching 8.6 billion, and continued minimal growth thereafter. Some analysts have questioned the sustainability of further world population growth, highlighting the growing pressures on the environment, global food supplies, and energy resources. While proven wrong in terms of benchmark dates, Malthusian predictions of calamitous outcomes for the human race surrounding overpopulation still persist.

Malthus

In his eighteenth century work *An Essay on the Principle of Population*, Malthus observed that an increase in a nation's food production improved the well-being of the populace, but the improvement was temporary because it caused population growth, which in turn restored the original per capita production level. In other words, mankind had a propensity to utilize abundance for population growth rather than for maintaining a high standard of living, a view that has become known as the "Malthusian trap" or the "Malthusian spectre".

Malthus preferred that populations had a tendency to grow until the lower class suffered hardship and want resulting in greater susceptibility to famine and disease, a view that is sometimes referred to as a Malthusian catastrophe. Malthus wrote in opposition to the popular view in 18th-century Europe that saw society as improving and in principle as perfectible. He saw population growth as being inevitable whenever conditions improved, thereby precluding real progress towards a utopian society: "The power of population is indefinitely greater than the power in the earth to produce subsistence for man". As an Anglican cleric, Malthus saw this situation as divinely imposed to teach virtuous behaviour. Malthus wrote:

- That the increase of population is necessarily limited by the means of subsistence,
- That population does invariably increase when the means of subsistence increase, and,
- That the superior power of population is repressed by moral restraint, vice and misery.

Malthus criticized the Poor Laws for leading to inflation rather than improving the well-being of the poor. He supported taxes on grain imports (the Corn Laws), because food security was more important than maximizing wealth. His views became influential, and controversial, across economic, political, social and scientific thought. Pioneers of evolutionary biology read him, notably Charles Darwin and Alfred Russel Wallace.

Malthus proposes the principle that human populations grow exponentially (i.e., doubling with each cycle) while food production grows at an arithmetic rate (i.e. by the repeated addition of a uniform increment in each uniform interval of time). Thus, while food output was likely to increase in a series of twenty-five year intervals in the arithmetic progression 1, 2, 3, 4, 5, 6, 7, 8, 9, and so on, population was capable of increasing in the geometric progression 1, 2, 4, 8, 16, 32, 64, 128, 256, and so forth. This scenario of arithmetic food growth with simultaneous geometric human population growth predicted a future when humans would have no resources to survive on. To avoid such a catastrophe, Malthus urged controls on population growth.

On the basis of a hypothetical world population of one billion in the early nineteenth century and an adequate means of subsistence at that time, Malthus suggested that there was a potential for a population increase to 256 billion within 200 years but that the means of subsistence were only capable of being increased enough for nine billion to be fed at the level prevailing at the beginning of the period. He therefore considered that the population increase should be kept down to the level at which it could be supported by the operation of various checks on population growth, which he categorized as "preventive" and "positive" checks. He was wrong about both 'predictions'.

The chief preventive check envisaged by Malthus was that of "moral restraint", which was seen as a deliberate decision by men to refrain "from pursuing the dictate of nature in an early attachment to one woman", i.e. to marry later in life than had been usual and only at a stage when fully capable of supporting a family. This, it was anticipated, would give rise to smaller families and probably to fewer families, but Malthus was strongly opposed to birth control within marriage and did not suggest that parents should try to restrict the number of children born to them after their marriage. Malthus was clearly aware that problems might arise from the postponement of marriage to a later date, such as an increase in the number of illegitimate births, but considered that these problems were likely to be less serious than those caused by a continuation of rapid population increase.

He saw positive checks to population growth as being any causes that contributed to the shortening of human lifespans. He included in this category poor living and working conditions which might give rise to low resistance to disease, as well as more obvious factors such as disease itself, war, and famine. Some later writers modified his ideas, suggesting, for example, strong government action to ensure later marriages. Others did not accept the view that birth control should be forbidden after marriage, and one group in particular, called the Malthusian League, strongly argued the case for birth control, though this was contrary to the principles of conduct which Malthus himself advocated.

In terms of public policy, Malthus was a supporter of the protectionist Corn Laws from the end of the Napoleonic Wars. He emerged as the only economist of note to support duties on imported grain. Malthus argued, the Corn Laws would guarantee British self-sufficiency in food.

The Corn Laws were measures enforced in the United Kingdom between 1815 and 1846, which imposed restrictions and tariffs on imported grain. They were designed to keep grain prices high to favour domestic producers. The laws did indeed raise food prices and became the focus of opposition from urban groups who had far less political power than rural Britain. The Corn Laws imposed steep import duties, making it too expensive to import grain from abroad, even when food supplies were short. The laws were supported by Conservative landowners and opposed by Whig industrialists and workers. The Anti-Corn Law League was responsible for turning public and elite opinion against the laws, in a large, nationwide middle-class moral crusade with a Utopian vision.

The first two years of the Irish famine of 1845–1852 forced a resolution because of the urgent need for new food supplies. Prime Minister Sir Robert Peel, a Conservative, achieved repeal with the support of the Whigs in Parliament, overcoming the opposition of most of his own party. "Corn" included any grain that requires grinding, especially wheat. The laws were introduced by the Importation Act and repealed by the Importation Act 1846).

The economic issue was food prices. The price of grain was central to the price of the most important staple food, bread, and the working man spent much of his wages on bread. Malthus' criticism of the working class's tendency to reproduce rapidly, and his belief that this, rather than exploitation by capitalists, led to their poverty, brought widespread criticism of his theory.

Malthusians perceived ideas of charity to the poor, typified by Tory paternalism, were futile, as these would only result in increased numbers of the poor; these theories played into Whig economic ideas exemplified by the New Poor Laws of 1834. The Act was described by opponents as "a Malthusian bill designed to force the poor to emigrate, to work for lower wages, to live on a coarser sort of food", which initiated the construction of workhouses ('Houses of Refuge and Industry') despite riots, machine breaking and arson.

The political issue was a dispute between landowners (a long-established class, who were heavily over-represented in Parliament) and the new class of manufacturers and industrial workers (who were under-represented). The former desired to maximise their profits from agriculture by keeping the price at which they could sell their grain high. The latter wished to maximise their profits from manufacture by reducing the wages they paid to their factory workers—the difficulty being that men could not work in the factories if a factory wage was not enough to feed them and their families; hence, in practice, high grain prices kept factory wages high also. The Corn Laws enhanced the profits and political power associated with land ownership. Their abolition saw a significant increase of free trade. The great Malthusian dread was that "indiscriminate charity" would lead to exponential growth in the population in poverty, increased charges to the public purse to support this growing army of the dependent, and, eventually, the catastrophe of national bankruptcy. Though Malthusianism has since come to be identified with the issue of general over-population, the original Malthusian concern was more specifically with the fear of over-population by the dependent poor (throughout the British Empire).

Neo Malthusians

Neo-Malthusianism generally refers to people with the same basic concerns as Malthus, who advocates population control programmes, to ensure resources for current and future populations. In Britain the term *Malthusian* can also refer more specifically to arguments made in favour of preventive birth control, hence organizations such as the Malthusian League. Neo-Malthusians seem to differ from Malthus's theories mainly in their enthusiasm for contraception. Malthus, a devout Christian, believed that "self-control" (abstinence) was preferable to artificial birth control. In some editions of his essay, Malthus did allow that abstinence was unlikely to be effective on a wide scale, thus advocating the use of artificial means of birth control as a solution to population "pressure". Modern "neo-Malthusians" are generally more concerned than Malthus was, with environmental degradation and catastrophic famine than with poverty.

Many critics believe that the basis of Malthusian theory has been fundamentally discredited in the years since the publication of *Principle of Population*, often citing major advances in agricultural techniques and modern reductions in human fertility. Many modern proponents believe that the basic concept of population growth eventually outstripping resources is still fundamentally valid, and "positive checks" are still likely in humanity's future if there is no action to curb population growth. Malthusian terms can carry a pejorative connotation indicating excessive pessimism, misanthropy or inhumanity. Some proponents of Malthusian ideas believe that Malthus's theories have been widely misunderstood and misrepresented; these proponents believe his reputation for pessimism and inhumanity is ill deserved. Malthusian ideas have attracted criticism from a diverse range of differing schools of thought, including Marxists and socialists, libertarians and free market adherents, Roman

Catholics, Islamicists,, social conservatives, feminists and human rights advocates.

India is anticipated to overtake China as the world's most populous country by 2022. Human population control is the practice of intervening to alter the rate of population growth. Historically, human population control has been implemented by limiting a region's birth rate, by voluntary contraception or by government mandate. It has been undertaken as a response to factors including high or increasing levels of poverty, environmental concerns, and religious reasons. The use of abortion in some population control strategies has caused controversy, with religious organizations such as the Roman Catholic Church reacting, explicitly opposing any intervention in the human reproductive process. Other Religio-Cultural groupings (such as within Islamic factions) have also opposed family size controls.

The Precariat

The radical economist Guy Standing, whose hugely influential book The *Precariat* identified an emerging social class suffering the worst of job insecurity would most likely be attracted to rightwing populism – a tailored fit to the administration of newly elected U.S. president Donald Trump., His *The Precariat* has achieved cult status as the first account of this emerging class of people, facing lives of insecurity, moving in and out of jobs that give little meaning to their lives.

Standing warns that the rapid growth of the precariat is producing instabilities in society. It is a dangerous class because it is internally divided, leading to the villainisation of migrants and other vulnerable groups. And, lacking agency, its members may be susceptible to the siren calls of political extremism. He argues for a new politics, in which redistribution and income security are reconfigured and in which the fears and aspirations of the precariat are made central to a progressive strategy. Since the first edition of this book in 2011, the precariat has become an ever more significant global phenomenon, highly visible in the Occupy movement and in protest movements around the world. In a new preface Standing discusses such developments - are they indicative of the emergence of a new collective spirit, or do they simply reveal the growing size and growing anger of this new class?

Chapter 3 Longevity and Environmental Mitigants

Futurists and longevity activists, find themselves constantly in the position of describing to the graying population the impact of life extension in the near future, and the conversation almost always follows the same course. Most people around 50 years or older, sadly, have already accepted (and come to terms with) their "impending" mortality. They don't necessarily "want" to live longer and seem to accept the world as is.

Expanding upon the ideas and theories about both the future and coming medical advances, they start to see the possibility: Maybe they could live a radically longer life. They almost always inevitably say, "but the planet can't handle that many people."

They suddenly actually care about the planet's future, if only for a fleeting second when they think that they might live long enough to exist in it. Introduction of the concept of Abundance as introduced by Peter Diamandis and other futurists, that lays out a possible path for our continued existence on this planet. *Abundance: The Future Is Better Than You Think*

Green Revolutions

is a book by Peter H. Diamandis and Steven Kotler that was published in 2012. The writers refer to the book's title as being a future where nine billion people have access to clean water, food, energy, health care, education, and everything else that is necessary for a first world standard of living, thanks to technological innovation. The book was a commercial success. It debuted at #1 on both Amazon.com and Barnes & Noble's bestseller lists. Praise appeared in various publications such as Time and The Washington Post.

Looking at the below image, you can understand why they think this way:

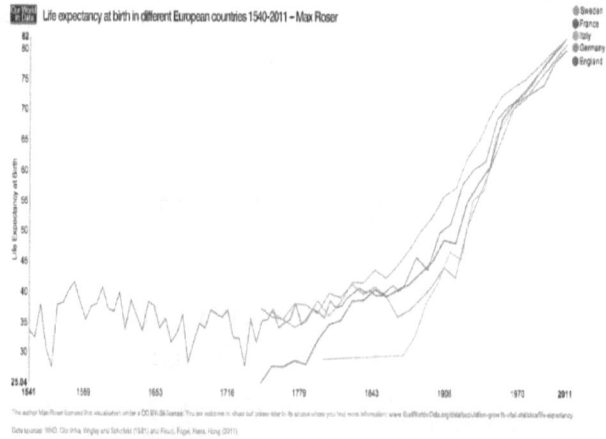

Life Expectancy Is Increasing Rapidly Over Time and Continuing To Increase

Green Revolutions

The writers refer to the book's title as being a future where nine billion people have access to clean water, food, energy, health care, education, and everything else that is necessary for a first world standard of living, thanks to technological innovation.

The book was a commercial success. It debuted at #1 on both Amazon.com and Barnes & Noble's bestseller lists. Positive reviews appeared in various publications such as Time and The Washington Post

The book's four main points are:

1. Technologies in computing, energy, medicine and many other areas are improving at an exponential rate and will soon enable breakthroughs that today seem impossible.
2. These technologies have allowed independent innovators to achieve startling advances in many areas of technology with little money or manpower. This is primarily achieved through incentive prize competitions.
3. Technology has created a generation of "techno-philanthropists" (such as Bill Gates) who are using their billions to try to solve seemingly unsolvable problems such as hunger and disease.

Green Revolutions

4. The lives of the world's poorest people are being improved substantially because of technology.

The book is divided into six parts:

- Perspective,
- Exponential Technologies,
- Building the Base of the Pyramid,
- The Forces of Abundance,
- Peak of the Pyramid, and
- Steering Faster.

It contains 19 chapters, a reference section with raw data, an appendix titled "Dangers of the Exponentials," and a Notes section for further reading.

Chapter 4 The Green Revolutions

The Green Revolution spread technologies that already existed, but had not been widely implemented outside industrialized nations. These technologies included modern irrigation projects, pesticides, synthetic nitrogen fertilizer and improved crop varieties developed through the conventional, science-based methods available at the time.

The 1st Green Revolution

The novel technological development of the first Green Revolution was the production of novel wheat cultivars. Agronomists bred cultivars of maize, wheat, and rice that are generally referred to as HYVs or "high-yielding varieties". HYVs have higher nitrogen-absorbing potential than other varieties. Since cereals that absorbed extra nitrogen would typically lodge, or fall over before harvest, semi-dwarfing genes were bred into their genomes. A Japanese dwarf wheat cultivar (Norin 10 wheat), which was sent to Washington, D.C. by Cecil Salmon, was instrumental in developing Green Revolution wheat cultivars. IR8, the first widely implemented HYV rice to be developed by IRRI, was created through a cross between an Indonesian variety named "Peta" and a Chinese variety named "Dee-geo-woo-gen".

With advances in molecular genetics, the mutant genes responsible for *Arabidopsis thaliana* genes wheat reduced-height genes (*Rht*) and a rice semidwarf gene were cloned. These were identified as gibberellin biosynthesis genes or cellular signaling component genes. Stem growth in the mutant background is significantly reduced leading to the dwarf phenotype. Photosynthetic investment in the stem is reduced dramatically as the shorter plants are inherently more stable mechanically. Assimilates become redirected to grain production, amplifying in particular the effect of chemical fertilizers on commercial yield.

HYVs significantly outperform traditional varieties in the presence of adequate irrigation, pesticides, and fertilizers. In the absence of these inputs, traditional varieties may outperform HYVs. Therefore, several authors have challenged the apparent superiority of HYVs not only compared to the traditional varieties alone, but by contrasting the monocultural system associated with HYVs with the polycultural system associated with traditional ones.

Cereal production more than doubled in developing nations between the years 1961–1985.

Yields of rice, maize, and wheat increased steadily during that period.]

While agricultural output increased as a result of that Green Revolution, the energy input to produce a crop has increased faster, so that the ratio of crops produced to energy input has decreased over time. Green Revolution techniques also heavily rely on chemical fertilizers, pesticides and herbicides and rely on machines, which as of 2014 rely on or are derived from crude oil, making agriculture increasingly reliant on crude oil extraction. Proponents of the 'Peak Oil theory' maintain that a future decline in oil and gas production will lead to a decline in food production or even a Malthusian catastrophe.

Effects

The effects of the Green Revolution on global food security are difficult to assess because of the complexities involved in food systems.

The world population has grown by about four billion since the beginning of the Green Revolution and many believe that, without the Revolution, there would have been greater famine and malnutrition. India saw annual wheat production rise from 10 million tons in the 1960s to 73 million in 2006. The average person in the developing world consumes roughly 25% more calories per day now than before the Green

Revolution Between 1950 and 1984, as the Green Revolution transformed agriculture around the globe, world grain production increased by about 160%.

The production increases fostered by the Green Revolution are often credited with having helped to avoid widespread famine, and for feeding billions of people.

There are also claims that the Green Revolution has decreased food security for a large number of people. One claim involves the shift of subsistence-oriented cropland to cropland oriented towards production of grain for export or animal feed. For example, the Green Revolution replaced much of the land used for pulses that fed Indian peasants for wheat, which did not make up a large portion of the peasant diet.

Criticisms

A main criticism of the effects of the Green Revolution is the rising costs for many small farmers using HYV seeds, with their associated demands of increased irrigation systems and pesticides. A case study is found in India, where farmers are planting cotton seeds capable of producing Bt toxin. A criticism regarding the Green Revolution are the effects regarding the widespread commercialization and market share of organisations, particularly of the phasing out of seed saving practices in favour of purchasing of seeds, and concerns regarding the financial affordability of the adoption of patented crops amongst farmers, particularly of those in the developing world. This can allow larger farms, even foreign owned farming operations, to buy up local small hold farms.

The 2nd Green Revolution

While the first Green Revolution, she notes, was mostly publicly-funded (by the Indian Government). The Second Green Revolutionis driven by private [and foreign] interest - notably multinationals like Monsanto. Ultimately, this is leading to corporate concentration and ownership of most of India's and North America's farmlands.

Some criticisms generally involve some variation of the Malthusian principle of population. Such concerns often revolve around the idea that the Green Revolution is unsustainable, and argue that humanity is now in a state of overpopulation or overshoot with regards to the sustainable carrying capacity and ecological demands on the Earth.

Although 36 million people die each year as a direct or indirect result of hunger and poor nutrition, Malthus's more extreme predictions have frequently failed to materialize. In 1798 Thomas Malthus made his prediction of impending famine. The world's population had doubled by 1923 and doubled again by 1973 without fulfilling Malthus's prediction. Malthusian Paul R. Ehrlich, in his 1968 book *The Population Bomb*, said that "India couldn't possibly feed two hundred million more people by 1980" and "Hundreds of millions of people will starve to death in spite of any crash programmes." Ehrlich's warnings failed to materialize when India became self-sustaining in cereal production in 1974 (six years later) as a result of the introduction of Norman Borlaug's dwarf wheat varieties.

Since supplies of oil and gas are essential to modern agriculture techniques, a fall in global oil supplies could cause spiking food prices in the coming decades

Mono-Culture

Some have challenged the value of the increased food production of Green Revolution agriculture. Miguel A. Altieri, (a pioneer of agroecology and peasant-advocate), writes that the comparison between traditional systems of agriculture and Green Revolution agriculture has been unfair, because Green Revolution agriculture produces monocultures of cereal grains, while traditional agriculture usually incorporates polycultures.

These monoculture crops are often used for export, feed for animals, or conversion into biofuel. According to Emile Frison of Bioversity International, the Green Revolution has also led to a change in dietary habits, as fewer people are affected by hunger and die from starvation, but many are affected by malnutrition such as iron or vitamin-A deficiencies. Frison further asserts that almost 60% of yearly deaths of children under age five in developing countries are related to malnutrition

High-yield rice (HYR), introduced since 1964 to poverty-ridden Asian countries, including the Philippines, was found to have inferior flavour and be more glutinous and less savory than their native varieties. This caused its price to be lower than the average market value.

In the Philippines the introduction of heavy pesticides to rice production, in the early part of the Green Revolution, poisoned and killed off fish and weedy green vegetables that traditionally coexisted in rice paddies. These were nutritious food sources for many poor Filipino farmers prior to the introduction of pesticides, further impacting the diets of locals

A major critic of the Green Revolution, U.S. investigative journalist Mark Dowie writes: "The primary objective of the programme was geopolitical: to provide food for the populace in undeveloped countries and so bring social stability and weaken the fomenting of communist insurgency".

There is significant evidence that the Green Revolution weakened socialist movements in many nations. In countries such as India, Mexico, and the Philippines, *technological solutions* were sought as an alternative to expanding *agrarian reform* initiatives, the latter of which were often linked to socialist politics.

The transition from traditional agriculture, in which inputs were generated on-farm, to Green Revolution agriculture, which required the purchase of inputs, led to the widespread establishment of rural credit institutions. Smaller farmers often went into debt, which in many cases results in a loss of their farmland. The increased level of mechanization on larger farms made possible by the Green Revolution removed a large source of employment from the rural economy. Because wealthier farmers had better access to credit and land, the Green Revolution increased class disparities, with the rich–poor gap widening as a result. Because some regions were able to adopt Green Revolution agriculture more readily than others (for political or geographical reasons), interregional economic disparities increased as well. Many small farmers are hurt by the dropping prices resulting from increased production overall. However, large-scale farming companies only account for less than 10% of the total farming capacity. This is a criticism held by many small producers in the food sovereignty movement.

The new economic difficulties of small holder farmers and landless farm workers led to increased rural-urban migration. The increase in food production led to a cheaper food for urban dwellers, and the increase in urban population increased the potential for industrialization.

In the most basic sense, the Green Revolution was a product of globalization as evidenced in the creation of international agricultural research centers that shared information, and with transnational funding from groups like the Rockefeller Foundation, Ford Foundation, and United States Agency for International Development (USAID).

Bio-Diversity

The spread of Green Revolution agriculture affected both agricultural biodiversity (or agrodiversity) and wild biodiversity There is little disagreement that the Green Revolution acted to reduce agricultural biodiversity, as it relied on just a few high-yield varieties of each crop.

This has led to concerns about the susceptibility of a food supply to pathogens that cannot be controlled by agrochemicals, as well as the permanent loss of many valuable genetic traits bred into traditional varieties over thousands of years. To address these concerns, massive seed banks such as Consultative Group on International Agricultural Research's (CGIAR) International Plant Genetic Resources Institute (now Bioversity International) have been established.

There are varying opinions about the effect of the Green Revolutions on indigenous biodiversity. One hypothesis speculates that by increasing production per unit of land area, agriculture will not need to expand into new, uncultivated areas to feed a growing human population. However, land degradation and soil nutrients depletion have forced farmers to clear up formerly forested areas in order to keep up with production. A counter-hypothesis speculates that biodiversity was sacrificed because traditional systems of agriculture that were displaced sometimes incorporated practices to preserve wild biodiversity, and because the Green Revolution expanded agricultural development into new areas where it was once unprofitable or too arid. For example, the development of wheat varieties tolerant to acid soil conditions with high aluminium content permitted the introduction of agriculture in sensitive Brazilian ecosystems .

Before the Green Revolution, other Brazilian ecosystems were also significantly damaged by human activity, such as the once 1st or 2nd main contributor to Brazilian megadiversity Atlantic Rainforest (about 95% after 2010) and the important xeric shrublands called Caatinga mainly in Northeastern Brazil (about 50% after the 2010s) — deforestation of the Caatinga biome is generally associated with greater risks of desertification). This also caused many animal species to suffer due to their damaged habitats.

Nevertheless, the world community has clearly acknowledged the negative aspects of agricultural expansion as the 1992 Rio Treaty, signed by 189 nations, has generated numerous national Biodiversity Action Plans which assign significant biodiversity loss to agriculture's expansion into new domains.

The Green Revolution has also been roundly criticized as an agricultural model relying on a few staple and market profitable crops, and pursing a model which introduced restricted biodiversity in Mexico. One of the critics against these techniques and the Green Revolution as a whole was Carl O. Sauer, a geography professor at the University of California, Berkeley. Sauer asserts that these techniques of plant breeding would result in negative effects on both a country's resources, and the culture.

High yield agriculture has dramatic effects on the amount of carbon cycling in the atmosphere. The way in which farms are grown, in tandem with the seasonal carbon cycling of various crops, could possibly alter the impact carbon in the atmosphere has on global warming. Wheat, rice, and soybean crops, account for a significant amount of the increase in carbon in the atmosphere over the last 50 years.

Fossil Fuels and Pesticides

Most high intensity agricultural production is dramatically dependent on non-renewable resources. Agricultural machinery and transport, as well as the production of pesticides and nitrates all depend on fossil fuels. Moreover, the essential mineral nutrient phosphorus is often a limiting factor in crop cultivation, while phosphorus mines are rapidly being depleted worldwide. The failure to depart from these non-sustainable agricultural production methods could potentially lead to a large scale collapse of the current system of intensive food production within this century.

The consumption of the pesticides used to kill pests by humans in some cases may be increasing the likelihood of cancer in some of the rural villages using them. Poor farming practices including non-compliance to usage of masks and over-usage of the chemicals compound this situation. In 1989, WHO and UNEP estimated that there were around 1 million human pesticide poisonings annually. Some 20,000 (mostly in developing countries) ended in death, as a result of poor labeling, loose safety standards etc.

Long term exposure to pesticides such as organochlorines, creosote, and sulfate have been correlated with higher cancer rates and organochlorines DDT, chlordane, and lindane as tumor promoters in animals. Contradictory epidemiologic studies in humans have linked phenoxy acid herbicides or contaminants in them with soft tissue sarcoma (STS) and malignant lymphoma, organochlorine insecticides with STS, non-Hodgkin's lymphoma (NHL), leukemia, and, less consistently, with cancers of the lung and breast, organophosphorous compounds with NHL and leukemia, and triazine herbicides with ovarian cancer.

Bill Gates has been among the proponents of a second green revolution, saying: "Three quarters of the world's poorest people get their food and income by farming small plots of land...if we can make smallholder farming more productive and more profitable, we can have a massive impact on hunger and nutrition and poverty...the charge is clear—we have to develop crops that can grow in a drought; that can survive in a flood; that can resist pests and disease...we need higher yields on the same land in harsher weather."

. Gates made these remarks during the World Food Prize in 2009. He has made over 1.4 billion in philanthropic contributions towards agricultural developments.

Indian Experience

The Indian state of Punjab pioneered green revolution among the other states; transforming India into a food-surplus country. The state is witnessing serious consequences of intensive farming using chemicals and pesticide. A comprehensive study conducted by Post Graduate Institute of Medical Education and Research (PGIMER) has underlined the direct relationship between indiscriminate use of these chemicals and increased incidence of cancer in this region.

In 2009, under a Greenpeace Research Laboratories investigation, Dr Reyes Tirado, from the University of Exeter, UK conducted the study in 50 villages in Muktsar, Bathinda and Ludhiana districts revealed chemical, radiation and biological toxicity rampant in Punjab. Twenty percent of the sampled wells showed nitrate levels above the safety limit of 50 mg/l, established by WHO, the study connected it with high use of synthetic nitrogen fertilizers.

Norman Borlaug dismissed certain claims of critics, but also cautioned, "There are no miracles in agricultural production. Nor is there such a thing as a miracle variety of wheat, rice, or maize which can serve as an elixir to cure all ills of a stagnant, traditional agriculture." Of environmental lobbyists, he said:"some of the environmental lobbyists of the Western nations are the salt of the earth, but many of them are elitists. They've never experienced the physical sensation of hunger. They do their lobbying from comfortable office suites in Washington or Brussels...If they lived just one month amid the misery of the developing world, as I have for fifty years, they'd be crying out for tractors and fertilizer and irrigation canals.

Although the Green Revolution has been able to improve agricultural output in some regions in the world, there was and is still room for improvement. As a result, many organizations continue to invent new ways to improve the techniques already used in the Green Revolution. Frequently quoted inventions are the System of

Rice Intensification, b marker-assisted selection, agroecology, and applying existing technologies to agricultural problems of the developing world.

It is thought that genetic engineering of new crops and foods will take the lead in producing increased crop yield and nutrition, a third generation revolution in agriculture.

Chapter 5 Abundance Theorism

The idea of abundance is a guiding and comforting theory of the future because it lays out a path where we will not all be starving or fighting over resources, but where we will be living in a society of plenty, one that has solved the majors problems that we have created by focusing our economic principles of society so strongly on scarcity.

Abundance Theory is underpinned by energy as a root cause of our current situation. If we solve our energy scarcity problem, the energy can then be used to power the solutions that underlie almost all of our other problems, in particular agriculture.

Solar Power

Every day, enough solar energy falls on the world in a couple of hours to power us for the entire year. We are already actually tracking way ahead of almost every estimate on our amount of solar panel production and installation growth, and it seems to be speeding up due to exponential advances in technology. Unfortunately it continues to have an untenable infrastructure cost at this time.

As Elon Musk pointed out when launching the PowerWall, with today's solar panels (20% efficient) it would only take an area of solar panels the size of a few counties in Northwest Texas to power all of the US, while an area of panels the size of Spain could power the entire world.

The secret weapon of solar panels is actually batteries, and even batteries without solar panels, as they can smooth the production curve from conventional sources. With a battery system, one can store the energy produced in the middle of the night when demand is low and shift that usage out from during the middle of the day when ACs are blasting at full speed, factories are in high-production and every office is lit up.

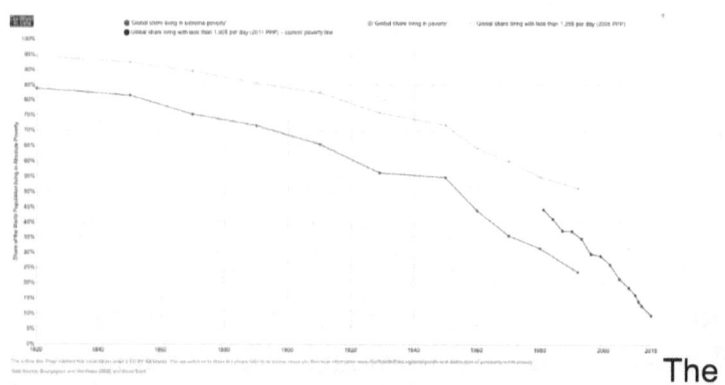

The percentage Of Global Population living in poverty is rapidly decreasing

Green Revolutions

Thorium

If solar panels for some reason don't generate enough energy, in the future we will very likely return to a form of nuclear power, which is fairly clean, especially if we are using thorium 233 instead of plutonium. Thorium-based nuclear power is nuclear reactor-based, fuelled primarily by the nuclear fission of the isotope uranium-233 produced from the fertile element thorium. According to proponents, a thorium fuel cycle offers several potential advantages over a uranium fuel cycle—including much greater abundance on Earth, superior physical and nuclear fuel properties, and reduced nuclear waste production.

Development of thorium power has significant start-up costs. Proponents also cite the lack of weaponization potential as an advantage of thorium, while critics say that development of breeder reactors in general (including thorium reactors, which are breeders by nature) increases proliferation concerns. Since about 2008, nuclear energy experts have become more interested in thorium to supply nuclear fuel in place of uranium to generate nuclear power. This renewed interest has been highlighted in a number of scientific conferences, the latest of which, ThEC13 was held at CERN by iThEC and attracted over 200 scientists from 32 countries.

A nuclear reactor consumes certain specific fissile isotopes to produce energy. The three most practical types of nuclear reactor fuel are:

- Uranium-235, purified (i.e. "enriched") by reducing the amount of uranium-238 in natural mined uranium. Most nuclear power has been generated using low-enriched uranium (LEU), whereas high-enriched uranium (HEU) is necessary for weapons.
- Plutonium-239, transmuted from uranium-238 obtained from natural mined uranium. Plutonium is also used for weapons.
- Uranium-233, transmuted from thorium-232, derived from natural mined thorium. This is the subject of this article.

Some believe thorium is key to developing a new generation of cleaner, safer nuclear power. According to an opinion piece by a group of scientists at the Georgia Institute of Technology, considering its overall potential, thorium-based power "can mean a 1000+ year solution or a quality low-carbon bridge to truly sustainable energy sources solving a huge portion of mankind's negative environmental impact."

After studying the feasibility of using thorium, nuclear scientists Ralph W. Moir and Edward Teller suggested that thorium nuclear research should be restarted after a three-decade shutdown and that a small prototype plant should be built. Research and development of thorium-based nuclear reactors, primarily the liquid fluoride thorium reactor (LFTR), a molten salt reactor (MSR) design, has been or is now being done in India, China, Norway, the United States, Israel and Russia.

The World Nuclear Association explains some of the possible benefits:"The thorium fuel cycle offers enormous energy security benefits in the long-term – due to its potential for being a self-sustaining fuel without the need for fast neutron reactors. It is therefore an important and potentially viable technology that seems able to contribute to building credible, long-term nuclear energy scenarios." Moir and Teller agree, noting that the possible advantages of thorium include "utilization of an abundant fuel, inaccessibility of that fuel to terrorists or for diversion to weapons use, together with good economics and safety features ... "Thorium is considered the "most abundant, most readily available, cleanest, and safest energy source on Earth," adds science writer Richard Martin.

Existing Canadian CANDU reactors can utilize Thorium 233 as fuel, The CANDU, for Canada Deuterium Uranium, is a Canadian-developed, pressurized heavy water reactor used to generate electric power. The acronym refers to its deuterium oxide (heavy water) moderator and its use of (originally, natural) uranium fuel. CANDU reactors were first developed in the late 1950s and 1960s by a partnership between Atomic Energy of Canada Limited (AECL), the Hydro-Electric Power Commission of Ontario, Canadian General Electric, and other companies.

There have been two major types of CANDU reactors, the original design of around 500 MWe that was intended to be used in multi-reactor installations in large plants, and the rationalized CANDU6 in the 600 MWe class that is designed to be used in single stand-alone units or in small multi-unit plants. CANDU6 units were built in Quebec and New Brunswick, as well as Pakistan, Argentina, South Korea, Romania, and China. Today there are 29 CANDU reactors in use around the world, and 13 "CANDU-derivatives" in India, developed from the CANDU design. After India detonated a nuclear bomb in 1974, Canada stopped nuclear dealings with India

Beyond that, we are working on some more "theoretical" technologies like actual fusion. The biggest collaborative project in the world right now is something called ITER, also known as "a star in a bottle" and it is an amazing feat of engineering, that if it works, will create a giant spherical magnetic field, inside of which will be a form of self-perpetuating fusion that is millions of degrees hot and is basically a star. And if that doesn't work, we have many other technologies in progress. What is more important to me than the actual method we use to solve it, is what unlimited energy or a level near that, will allow for.

Non-Traditional Farming

The world can use that water to grow our food, but it also turns out that growing food outside may not actually be the best way to do it. Sunlight is transient and even the best land in the world only gets 8-10 hours or so of desirable light per day, and then you have insects, parasites and a host of other shortcomings of traditional agriculture. Not to mention that most places are not ideal for growing food due to weather, topography, temperature, soil, etc. Farms are also generally not located near where people live, which causes logistical issues if the food will spoil, etc.

The solution, which is already being implemented today, is something called vertical farming. The first large-scale vertical farm was launched in Japan near where the Fukushima nuclear reactor had caused significant radiation (I completely recognize the irony of this, based on my statements about nuclear power above). The project was meant to show that vertical farms can be used to grow food, even in inhospitable areas. And we actually see the next ones being installed in places where there are land shortages (Singapore) and where growing conditions are not desirable and/or traditional farming is impossible (Nepal & Tibet).

Vertical farming is pretty much exactly what it sounds like, growing food indoors and inside buildings where rows of plants can be stacked on top of each other. A single large building can grow food for the surrounding neighborhood, and it only has to travel mere feet to its destination.

The further benefits of vertical farming are that it takes significantly less water than conventional agriculture because the plants grow in an environment that has no soil, no bugs, and no evaporation. On top of that, using specially designed LEDs that have the specific wavelengths that plants need for growth (red and blue), they can simulate 12-18 hours of sunlight per day and grow plants much more rapidly.

The farm in Japan already mentioned is growing over 10,000 heads of lettuce in 25,000 square feet, so it is 100x more productive than traditional farming, while using 80% less food waste, 40% less energy and 99% less water.

And mind you, this is all organic, no pesticide, and hyperlocal. It is truly the future of agriculture. Right now they can only grow leafy greens, but they are working on bigger and more hearty plants. These vertical farms are popping up all over the US and the largest one in the world is being built in New Jersey, just across from New York City.

So now that we have energy, water, and food resolved; futurists deal with space on our planet. So it turns out that most of the planet and the interior of the US is largely uninhabited by humans, and we actually have plenty of room, we just don't utilize it. If everyone in the world lived at the same population density as Paris or New York, all 7-8 billion of us could actually fit in an area the size of Texas.

3-D Printing

We can easily provide enough homes for anyone who wants them, and I imagine in the future we will print entire cities for those who do not have a place to live. 3D printing, also known as additive manufacturing (AM), refers to processes used to synthesize a three-dimensional object in which successive layers of material are formed under computer control to create an object. Objects can be of almost any shape or geometry and are produced using digital model data from a 3D model or another electronic data source such as an Additive Manufacturing File (AMF) file.

Futurist Jeremy Rifkin claimed that 3D printing [AM] signals the beginning of a third industrial revolution, succeeding the production line assembly that dominated manufacturing starting in the late 19th century. The term 3D printing's origin sense is in reference to a process that deposits a binder material onto a powder bed with inkjet printer heads layer by layer. More recently, the term is being used in popular vernacular to encompass a wider variety of additive manufacturing techniques. United States and global Technical standards use the official term *additive manufacturing* for this broader sense. ISO/ASTM52900-15 defines seven categories of AM processes within its meaning:

- Binder Jetting,
- Directed Energy Deposition,
- Material Extrusion,
- Material Jetting,
- Powder Bed Fusion,
- Sheet Lamination and
- Vat Photopolymerization

Chapter 6 Farming and 3D Printing

The 3rd Green Revolution is 3-D Printing, Borne out of necessity, farmers possess an amazing variety of mechanical skills. In the years to come, those skill sets are set to expand—particularly in the field of 3D printing. As prices fall and efficiency rises, 3D printers might become standard tools on the farm.

3D printing could empower farmers to increase self-sufficiency. The technology might reach a point where machinery manufacturers allow farmers to purchase design files online and print products themselves. Or, local dealers with 3D printing capability could make transitional parts for farmers to use while waiting for permanent replacements [and conceivably have long life on their own].

Ben Bernard, computer service specialist, North Dakota State University, is cautiously optimistic about 3D printing benefiting agriculture. "Farmers usually live in a rural setting and need custom parts. Along with welding or electrical capabilities, 3D printing capability might become normal for a farming operation."

Bernard says the agriculture industry should be excited about 3D printing materials—particularly polylactic acid or PLA. Made from corn starch, PLA offers another market possibility for corn producers. Older 3D printing technologies can cost $5 to $10 per cubic inch, but PLA is only 25¢ to 50¢ per cubic inch [up to a 98% cost saving].

Attempts have been made to have the North Dakota Corn Growers Association fund PLA research because suddenly there's a lot of PLA demand. Bernard says. "Can we derive a filament even stronger than what we've got and find new markets for ag products?"

Nathan Hulstein, president of GVL Poly, Litchfield, Minn., says: "3D printing will continue to gain popularity, especially in product development for manufacturers. "In the distant future, I think individual farmers may print their own replacement parts, but in the next decade, ag engineers will use 3D to speed up design cycles," he says. GVL Poly is centered on season-related agricultural products, and field testing opportunities are restricted to a narrow time frame. When a prototype design fails, a year may pass before testing opportunities reopen".

However, 3D printing allows manufacturers to change prototypes multiple times within a testing cycle. Rather than taking four or five expensive mold design changes, 3D printing allows for a single mold to be built with no design changes—an exponential level of improvement. The systems may require three or four mouldings to replace as many as eighty separate machined components requiring assembly.

"Farmers should get excited because machinery manufacturers have increasing opportunity to try more options during test season," Hulstein says. "This allows farmers to get access to machinery improvements quicker as products hit the market at a faster rate."

3D printing offers a major savings in tooling, says Rye DeGarmo, AGCO engineering manager for seeding and tillage. A tooled cast part can cost between $8,000 and $12,000—with the potential of a 10-week wait. With 3D printing, DeGarmo can instead make 10 iterations for testing before committing to tooling costs. Even today the total capital and constituent material costs for a high-end printer to do this would be half the cost of the ordered 'conventional part'. Printing the part would require six to twelve hours compared to a ten week down time for conventional part delivery:"

The 3-D Farm-Bot

Green Revolutions

Entrepreneur Rory Aronson has developed the world's first open-source CNC farming machine. The FarmBot Genesis, available to pre-order as a kit from July, is made from 3D printable plastic components, and can be used to remotely plant, water, and monitor a garden.

Apparently everyone's a farmer is the premise behind FarmBot Genesis, a brilliant idea from entrepreneur Rory Aronson and his small California-based startup, which has built a 3D printed CNC farming machine, 100% open source, which can be controlled digitally through a simple web-based interface.

Aronson's vision is to make precision agriculture open and accessible to all, and his FarmBot project addresses this. The FarmBot Genesis system, the first scalable iteration of the concept, can be built for around $1,500 to $4,000, and features an intuitive, game-like interface through which users can tailor their patch to their liking. Maintenance sequences can be easily scheduled, allowing complete remote control over the miniature farm. Need to water different seeds at different times? Simply pop the time into the schedule, and the FarmBot hardware will do its thing using linear guides in the X, Y, and Z directions. The tools of the machine can even be controlled manually in real time.

Green Revolutions

"After graduating with a degree in mechanical engineering, I decided I wanted to reinvent the way food is grown in order to adapt to the growing use of technology in people's lives," Aronson said. "When I thought about rebuilding the agricultural process, a robot is what I pictured. It is quite literally a robotic device that someone can control through their smartphone or laptop. It's simple enough to use in your home but sophisticated enough to adapt to a larger scale."

Although building a garden-size CNC machine might sound like a massive task, Aronson has endeavored to make the project as simple as possible, by providing both full documentation for building your own and providing the entire FarmBot as a kit. At its core it powered by a Raspberry Pi 3, an Arduino Mega 2560, and a RAMPS 1.4 shield, all easily affordable. All of the digital farm's plastic components can be made with a 3D printer, and the flat connecting plates required for the project can be made with a waterjet, plasma, or laser cutter; a CNC mill [or a hacksaw and drill press].

In order to cultivate whatever plants and vegetables you might choose to grow with FarmBot, Aronson has designed a number of digitally controllable tools which function just like their manual equivalents. Seed injection, watering, and weed suppression are all made simple with dedicated attachments, while the open-source nature of the project allows digital gardeners to create whatever extra tools they need before (hopefully) sharing them with the FarmBot community.

There are so many clever features devised for the FarmBot project. The system is able to gather and use real-time weather data, so it knows when watering is necessary and when it can let nature do the work. It can also store this rainwater in a barrel and release it at programmed times, and can even power itself with solar panels. The project is, of course, in its early stages, but there are endless possibilities for FarmBot, and only time will tell how users will put the system to use.

"Right now, we're just touching the tip of the iceberg as far as 3D capabilities for agriculture. As this becomes more common, we'll use 3D all the more, but prices have to drop," he says. DeGarmo cites costs as the main impediment keeping 3D printing from going mainstream. A factory churning out 1,000 items in a perfected, streamlined process compared to a 3D printer producing a single, one-off part equates to a major difference in costs.

Costing for Farms

"It's fascinating to me because there is real potential for 3D printing technology to help individual farmers and not just factories," DeGarmo notes. "The cost of materials may be the driver, and if they go down, there are a lot of farm parts that would get made on 3D printers. The cost breakover will determine how fast that happens in the future."

Based on extensive evaluations and hours of testing of more than a dozen models in different price ranges, a top overall pick for those on a budget is the XYZ da Vinci Mini ($270), as its auto-calibration features are helpful to novices and it produces good prints at decent speeds. (If you want to save a little money, consider the da Vinci miniMaker, which is selling for $45 less than the Mini on Amazon, but is essentially the same printer. It connects via USB instead of wirelessly like the Mini.)

Those looking to print in a variety of materials should check out the LulzBot Mini ($1,250), which supports ABS, nylon, polycarbonate and polystyrene. 3D-printing enthusiasts and professional designs will appreciate the two swappable extruders and excellent print quality of the Ultimaker 3 ($3,495).

Because 3D printer makers are constantly announcing new models that cost less, print things faster and produce larger objects than ever before, it's worth knowing about just-released printers before you buy. Check out our New & Notable section below for the latest information on newly released printers, including some of the more eye-catching models we spotted during January's CES event in Las Vegas.

To make it easier to know which 3D printer is right for you, here are a few things to look out for, along with more information on all of our top picks.

- Printer type: There are two main types of 3D printers: FFM (fused filament manufacturing) and SLA (stereo lithography). FFM printers work by melting a plastic filament in a moving printhead to form the model. SLA printers use an ultraviolet (UV) laser to solidify a resin, focusing the laser to form the solid model. FFM printers are generally cheaper, simpler and easier to use, although SLA models like the XYZprinting Nobel 1.0 (around $1,000) are lowering the price difference.
- Printing materials: Whichever type of printer you choose, pay attention to the type of material it can use to print. The filament material used by FFM printers like

the LulzBot TAZ 6 is available in several different materials, such as PLA (a brittle, biodegradable material), ABS (the same plastic used in Lego blocks), nylon, TPE (a soft, rubberlike material) and HDPE light, tough polystyrene). Many of these materials, particularly PLA and ABS, are available in a huge range of colors. Filaments come in two sizes: 1.75 mm and 3 mm, which are not interchangeable.

SLA printers have fewer options than their FFM counterparts, but printers like the Form 2 can use resins that produce models ranging from very rigid to flexible and rubbery. The best printers can use a wide range of materials, each of which comes with its own strengths and weaknesses. (HDPE, for example, is light and tough, but not suitable for food use, while nylon is food-safe.)

Note that, some printers only allow the use of approved materials or materials produced by the same company that made the printer. In that sense, those types of 3D printers are like more traditional paper printers: The manufacturers sell the hardware cheaply and then make money back on the consumables. (Our top budget 3D printer, the da Vinci Mini, only works with PLA filament from manufacturer XYZprinting, for example;

however XYZ's filament costs about the same as most third-party materials.) Other 3D printers place no restrictions on the type or origin of the material.

Print volume: All printers have limits on the size of the 3D print they can produce. That limit is defined by the size of the print bed and how far the printer can move the printhead. This is usually measured in cubic inches, but you should also pay attention to each of the individual dimensions, which determine the maximum size 3D print the device can create. So, for example, if a printer like the LulzBot Mini has a print volume of 223 cubic inches (6.2 x 6 x 6 inches), it can print objects that are up to just less than 6 inches high, wide and deep.

Print speed and quality: 3D printing is a slow business, and at present, there's no way to get around this. You should expect a 3- to 4-inch model to typically take between 6 and 12 hours to print, depending on the print quality you select. That's because of the way 3D printing works: The print is constructed in layers. The thicker these layers are, the quicker the print is produced but the lower the print quality is, as the layers become more visible. So, there is a trade-off between print speed and print quality.

The best printers will allow you to determine which way you want to go with this, producing prints quickly or more slowly but at higher quality. The best printers offer a wide range of quality settings, from fast (but low quality) to slow (but high quality).

XYZprinting puts 3D printing within anyone's budget with this $270 device. Easy-to-use software and an auto-calibration feature make the da Vinci Mini a great 3D printer for novices, and you'll get clean, smooth prints at decent speeds. (Printing at higher resolutions slows things down considerably, though.) You're restricted to using PLA filament from XYZ, but since that filament isn't expensive, that's a fair compromise to make for an attractively priced 3D printer with good-looking output.

The Lulzbot Mini offers big performance in a small package, complete with carrying handle. It offered high-quality prints in our tests and it can print in several materials beyond the standard PLA and ABS plastic. There is also a high temperature extruder for using such materials as Nylon and metal-filled plastic filament. The Lulzbot Mini supports only a single extruder, though, so you have to manually swap the filament out if you want to use more than one color or type in a print job. The user-friendly Cura software makes your projects easy to execute.

You'll pay a comparably steep price for Ultimaker's latest $3500 printer, but if you're a design professional or serious 3D-printing user, the Ultimaker 3 is more than worth the cost. Print quality is excellent — some of the best we've seen from a 3D printer, even in draft mode — and the Ultimaker 3 supports a wide range of materials. Excellent software makes it easy to manage prints, and a redesigned printhead with two swappable extruders adds to the Ultimaker 3's impressive flexibility.

The latest SLA printer from Formlabs improves on its predecessor by expanding the print area to 224 cubic inches, up from 156 cubic inches on the Form 1+. An updated print mechanism means greater print consistency, and the Form 2 can handle third-party resins — both improvements over past versions. Most importantly, though, the Form 2 produces excellent prints featuring fine detail with clean, sharp edges. Its price tag puts it out of the reach of casual users, but engineers, artists and jewelers will appreciate the Form 2's performance.

Construction 3-D Printing

Green Revolutions

Construction 3D printing refers to various technologies that use 3D printing as a core method to fabricate buildings or construction components, a different technology of great importance to the agricultural community where infra-structure capital cost of out buildings and other structures have skyrocketed for conventional 'rural' construction techniques and materials

Current machines are being integrated into automated and semi automated production lines and, because of the scale of construction, will feature elements of additive, subtractive and formative manufacturing processes, to handle material deposition at one scale and finishing at another. Because of the cost, 3D printing at construction scales demands clever design and can respond to the demands of architects and engineers for high value, high performance building components. Potential advantages of these technologies include faster construction, lower labour costs, increased complexity and/or accuracy, greater integration of function and less waste produced. There are a variety of 3D printing methods used at construction scale, these include the following main methods:

- extrusion (concrete/cement, wax, foam, polymers),

- powder bonding (polymer bond, reactive bond, sintering) and
- additive welding.

3D printing at a construction scale will have a wide variety of applications within the private, commercial, industrial and public sectors. Development has been slow and sporadic, since its development in the mid 1990s, where initially it was explored as a scaled version of mainstream 3D printing, having both novelty value and early research funding in both the US and Europe. The term 'Construction 3D Printing' was first coined by James B Gardiner in 2011.

A number of different approaches have been demonstrated to date which include on-site and off-site fabrication of buildings and construction components, using industrial robots, gantry systems and tethered autonomous vehicles. Demonstrations of construction 3D printing technologies to date have included fabrication of housing, construction components (cladding and structural panels and columns), bridges, artificial reefs, follies and sculptures.

Current efforts focus on integrating the advantages of digital fabrication within factory based construction manufacturing. Stand alone and on-site machines are in planning and research, ranging from modified autonomous concrete/gypsum/mineral paste pumping/spraying, composite fiber spinning and ultimately swarm construction agents, where construction 3D printing merges with robotics and AI systems

Early construction 3D printing development and research have been under way since 1995. Two methods were invented, one by Joseph Pegna which was focused on a sand/cement forming technique which utilized steam to selectively bond the material in layers or solid parts: this technique was never demonstrated. The second technique, Contour Crafting by Behrohk Khoshnevis, initially began as a novel ceramic extrusion and shaping method, as an alternative to the emerging polymer and metal 3D printing techniques, and was patented in 1995. Khoshnevis realized that this technique could exceed these techniques where "current methods are limited to fabrication of part dimensions that are generally less than one meter in each dimension". Around 2000, Khoshnevis's team at USC Vertibi began to focus on construction scale 3D printing of cementitious and ceramic pastes, encompassing and exploring automated integration of modular reinforcement, built-in plumbing and electrical services, within one continuous build process.

In 2003, Rupert Soar secured funding and formed the freeform construction group at Loughborough University, UK, to explore the potential for up-scaling existing 3D printing techniques for construction applications. Early work identified the challenge of reaching any realistic break-even for the technology at the scale of construction and highlighted that there could be ways into the application by massively increasing the value proposition of integrated design (many functions, one component). In 2005, the group secured funding to build a large-scale construction 3D printing machine using 'off the shelf' components (concrete pumping, spray concrete, gantry system) to explore how complex such components could be and realistically meet the demands for construction.

In 2005 Enrico Dini, Italy, patented the D-Shape technique, employing a massively scaled powder jetting/bonding technique over an area approximately 5m x 5m x 2.5m. This technique although originally developed with epoxy resin bonding system was later adapted to use inorganic bonding agents. This technology has been used commercially for a range of projects in construction and other sectors including for [artificial reefs].

3D Concrete Printing began in 2008 at Loughborough University, UK, headed by Richard

Buswell ,moving from a gantry based technique to an industrial robot, which they succeeded in licensing the technology to Skanska in 2014.

From 2013 onward technologies of construction printing have advanced rapidly, and include:

- Winsun (Shanghai WinSun Decoration Design Engineering Co) launched their concrete printer on April Fools' Day 2014 with an article on 3ders. The company claimed to have printed 10 houses in a day each 200m2, however visual scaling from the photographs does not support the claim for the buildings size. The demonstration of a large number of 3D printed buildings however was the first of its kind and indicates the capability of the construction 3D printer developed. On January 18, 2015 the company gained further press coverage with the unveiling of 2 further buildings, a mansion style villa and a 5 storey tower, using 3D printed components. Detailed photographic inspection indicates that the buildings were fabricated with both precast and 3D printed components. The buildings stand as the first complete structures of their kind fabricated using construction 3D printing technologies. In May 2016 a new 'office building' was opened in Dubai. The 250-square-metre space (2,700 square foot) is what Dubai's Museum of the Future project is calling the world's first 3D-printed office building. Although Winsun claim to have been working on 3D printing for many years, no publicly available information has been found to support the claim.

- FreeFAB] Wax™, invented by James B Gardiner and Steven Janssen at Laing O'Rourke (Construction Company). The patented technology has been in development since March 2013. The technique uses construction scale 3D printing to print high volumes of engineered wax (up to 400L/hr) to fabricate a 'fast and dirty' 3D printed mould for precast concrete, glass fibre reinforced concrete (GRC) and other sprayable/cast-able materials. The mould casting surface is then 5 axis milled removing approximately 5mm of wax to create a high quality mould (approximately 20 micron surface roughness). After the component has cured, the mould is then either crushed or melted-off and the wax filtered and re-used, significantly reducing waste compared to conventional mould technologies. The benefits of the technology are fast mould fabrication speeds, increased production efficiencies, reduced labour and virtual elimination of waste by re-use of materials for bespoke moulds compared to conventional mould technologies. The system was originally demonstrated in 2014 using an industrial robot. The system was later adapted to integrate with a 5 axis high speed gantry to achieve the high speed and surface milling tolerances required for the system. The first industrialised system is installed at a Laing O'Rourke factory in the United Kingdom and is due to start industrial production for a prominent London project in late 2016.]

- BetAbram is a simple gantry based concrete extrusion 3D printer developed in Slovenia. This system is available commercially, offering 3 models (P3, P2 and P1) to consumers since 2013. The largest P1 can print objects up to 16m x 9m x 2.5m (sizes common to 'rooms' within residences).

- Dutch architect Janjaap Ruijssenaars's performative architecture 3D-printed building was planned to be built by a partnership of Dutch companies. The house was planned to be built in the end of 2014, but this deadline wasn't met. The companies said that they are still 100% sure the house will be printed.

Various approaches to Construction 3D Printing are being researched. Two of these are Contour crafting and D-Shape. Other approaches involve direct sintering of inorganic raw materials to build composite ceramic building structures, similar to the approach used with metals in direct metal laser sintering.

In the Netherlands, DUS Architects is 3D printing a 3D Printed Canal House, together with an international team of partners. The 3D Print Canal House links science, design, construction and community at an open building site in the heart of Amsterdam. Their aim is to demonstrate how 3D printing could revolutionize construction by increasing efficiency and reducing pollution and waste, and offer new tailor made housing solutions worldwide. 3D printing could also play a significant role in the quick build of low-cost housing in impoverished areas and those affected by disasters. The 3D Print Canal House is currently under construction at a canal-side plot in Amsterdam – an open 'expo-site' that it is proving to be a popular visitor attraction for the public. At the heart of the site, is the Kamermaker, or Room Builder – which is essentially a scaled-up version of a table-top 3D printer.

The Kamermaker prints building blocks from molten bio-plastic. This is currently a mix of 80% plant oil reinforced with microfibers, although this formula is still under development with the project's materials partner Henkel. For reinforcement, the blocks have an internal honeycombed centre that can be back-filled with Eco concrete. It also provides space for pipes, wiring and data cables to be installed internally.

The building blocks are then used to form component parts that can be slotted together like Lego to create a 4-storey, 13-room structure modelled on a traditional Dutch canal house. One of the most distinct design features of the Canal

House is its geometrically faceted plastic façade. 3D Print House Building Blocks This gives a contemporary 3D print twist to the traditional canal house silhouette. The ability to print ornamental detailing on demand is a key design benefit of 3D modelling and printing in the building industry. With costly labour-intensive work reduced, custom-designed homes would become more accessible. So what are the main benefits of printing a house? Waste materials are a big problem for the building industry, but with 3D printing only the necessary raw materials are produced for each project. An added bonus is that 3D printer 'ink' can be made from recycled plastic waste. If printing on site, transport costs and CO_2 emissions are greatly reduced – as are dust and noise levels. And when the building is no longer needed, it can be shredded and recycled. Another key driver for developing this technology within the construction industry is the growing need for rapidly produced housing. In this respect, 3D printing has the potential to reshape the way in which we build our cities – especially as Megacities are on the increase around the globe. The 3D Print Canal House was the first full-scale construction project of its kind to get off the ground. In just a short space of time, the Kamermaker has been further developed to increase its production speed by 300%. However, progress has not been swift enough to claim the title of 'World's First 3D Printed House'.

Dutch and Chinese demonstration projects are slowly constructing 3D-printed buildings, using the effort to educate the public to the possibilities of the new plant-based building technology and to spur greater innovation in 3D printing of residential buildings.

Claims have been made by Behrokh Khoshnevis since 2006 for 3D printing a house in a day, with further claims to notionally complete the building in approximately 20 hours of "printer" time. By January 2013, working versions of 3D-printing building technology were printing 2 metres (6 ft 7 in) of building material per hour, with a follow-on generation of printers proposed to be capable of 3.5 metres (11 ft) per hour, sufficient to complete a building in a week. The Chinese company WinSun has built several houses using large 3D printers using a mixture of quick drying cement and recycled raw materials. Ten demo houses were built in 24 hours, each costing US$5000.

We should not live in a society where people have to live on the street and or go to bed hungry. We could solve these problems today if we actually tried, but we really do not put the resources or the right mentality towards it. This is something that will come naturally as we will run out of excuses to not do it and to look at ourselves in the mirror and wonder why we allow society to function in that way.

Climate Change and Farm-Abundance

The 2016 Paris talks may be one of the first real steps to limiting CO2 production in the world, but the problem is that even if the best proposals currently being put forward are adopted, it will likely not be enough. It has been determined that we need to stop the climate from rising more than 2 degrees, or we will see dramatic changes that will *not* be good for our continued existence. This is where we need to call on technology to save us, and this is the pending question of the future: Can we use technology to solve the problems that technology has caused in the first place?

We must invent the technologies to not only stop the output of greenhouse gasses, but to reverse them? This is known as geoengineering, and it holds the true promise to fix the problems we have created.

There are proposals to have autonomous ships sailing around the ocean that spray salt water, sulfur dioxide, or other particles that can reflect light back into the atmosphere or absorb some CO2 so the planet does not warm. There are other devices like filtrations systems that can pull carbon dioxide out of the air (like trees do), and sequester it, so that we can reduce the existing levels. A company called CarbonEngineering is already doing this.

This 'abundance' is the future for us, a future where we spend our resources on fixing the planet and bringing it back to a pristine state. The jobs of the future will likely be related to restoring the planet, re-seeding the world's biodiversity, restocking our oceans, and making our air breathable again. When people can stop having wars over resources, fighting for oil or water or food, we can bring humanity together to focus on the big picture: Our continued survival as a whole.

This is where the power of living significantly longer has an impact. It makes people actually think about the future. It forces us, as a society, to come up with a solution for the problems that could compromise our future.

Abundance gives people hope for a world where people actually want to be alive. By brightening the prospects of the future, we brighten the prospects for humanity today. We stop looking at this planet as a temporal place, where one exists and then dies, nod we start looking at it as an integrated system and how we will exist within it. If you throw away that plastic cup and it is going to be around for the next 500 years with you, then you may think twice about having that cup created in the first place. If we extend the vision of people out into the centuries timeframe, then we are forced to deal with the truly pressing issues of our time that are otherwise looked at as "my grandchildren's problem," and the issues then become all of our problem.

And when all of us are "selfishly" focused on making sure our existence is sustainable and the future is bright, then we will truly be on the right path and actually ensure that abundance occurs, because it really is the only possible way we can survive on this planet.

Chapter 7 Protein Projections

There are three major sources of protein: plant, livestock and aquaculture (fish and shellfish).

Consumption of fruits and vegetables plays a vital role in providing a diversified and nutritious diet. A low consumption of fruits and vegetables in many regions of the developing world is, however, a persistent phenomenon, confirmed by the findings of food consumption surveys. Nationally representative surveys in India), for example, indicate a steady level of consumption of only 120-140 g per capita per day, with about another 100 g per capita coming from roots and tubers, and some 40 g per capita from pulses. This may not be true for urban populations in India, who have rising incomes and greater access to a diverse and varied diet. In contrast, in China, - a country that is undergoing rapid economic growth and transition - the amount of fruits and vegetables consumed has increased to 369 g per capita per day by 1992.

At present, only a small and negligible minority of the world's population consumes the generally recommended high average intake of fruits and vegetables. In 1998, only 6 of the 14 WHO regions had an availability of fruits and vegetables equal to or greater than the earlier recommended intake of 400 g per capita per day. The relatively favourable situation in 1998 appears to have evolved from a markedly less favourable position in previous years, as evidenced by the great increase in vegetable availability recorded between 1990 and 1998 for most of the regions. In contrast, the availability of fruit generally decreased between 1990 and 1998 in most regions of the world.

The increase in urbanization globally is another challenge. Increasing urbanization will distance more people from primary food production, and in turn have a negative impact on both the availability of a varied and nutritious diet with enough fruits and vegetables, and the access of the urban poor to such a diet. Nevertheless, it may facilitate the achievement of other goals, as those who can afford it can have better access to a diverse and varied diet. Investment in periurban horticulture may provide an opportunity to increase the availability and consumption of a healthy diet.

Global trends in the production and supply of vegetables indicate that the current production and consumption vary widely among regions. It should be noted that the production of wild and indigenous vegetables is not taken into account in production statistics and might therefore be underestimated in consumption statistics. In 2000, the global annual average per capita vegetable supply was 102 kg, with the highest level in Asia (116 kg), and the lowest levels in South America (48 kg) and Africa (52 kg). These figures also include the large amount of horticultural produce that is consumed on the farm.

In recent years the growth rates of world agricultural production and crop yields have slowed. This has raised fears that the world may not be able to grow enough food and other commodities to ensure that future populations are adequately fed. However, the slowdown has occurred not because of shortages of land or water but rather because demand for agricultural products has also slowed. This is mainly because world population growth rates have been declining since the late 1960s, and fairly high levels of food consumption per person are now being reached in many countries, beyond which further rises will be limited. It also true that, while diminishing, a high share of the world's population still remains in poverty and hence lacks the necessary income to translate its needs into effective demand. As a result, the growth in world demand for agricultural products is expected to fall from an average 2.2% per year over the past 30 years to an average 1.5% per year for the next 30 years. In developing countries the slowdown will be more dramatic, from 3.7% per year to 2% per year, partly as a result of China having passed the phase of rapid growth in its demand for food. Global food shortages are unlikely, but serious problems already exist at national and local levels, and may worsen unless focused efforts are made.

The annual growth rate of world demand for cereals has declined from 2.5% per year in the 1970s and 1.9% per year in the 1980s to only 1% per year in the 1990s. Annual cereal use per person (including animal feeds) peaked in the mid-1980s at 334 kg and has since fallen to 317 kg. The decline is not a cause for alarm, it is largely the natural result of slower population growth and shifts in human diets and animal feeds. During the 1990s, however, the decline was accentuated by a number of temporary factors, including serious economic recessions in the transition countries and in some East and South-East Asian countries.

The growth rate in the demand for cereals is expected to rise again to 1.4% per year up until 2015, slowing to 1.2% per year thereafter. In developing countries overall, cereal production is not expected to keep pace with demand. The net cereal deficits of these countries, which amounted to 103 million tonnes or 9% of consumption in 1997-1999, could rise to 265 million tonnes by 2030, when they will be 14% of consumption. This gap can be bridged by increased surpluses from traditional grain exporters, and by new exports from the transition countries, which are expected to shift from being net importers to being net exporters.

Oil crops have seen the fastest increase in area of any crop sector, expanding by 75 million hectares between the mid-1970s and the end of the 1990s, while cereal area fell by 28 million hectares over the same period. Future per capita consumption of oil crops is expected to rise more rapidly than that of cereals. These crops will account for 45 out of every 100 extra kilocalories added to average diets in developing countries between now and 2030.

There are three main sources of growth in crop production:

- expanding the land area,
- Increasing the frequency at which it is cropped (often through irrigation), and boosting yields.

It has been suggested that growth in crop production may be approaching the ceiling of what is possible in respect of all three sources. A detailed examination of production potentials does not support this view at the global level, although in some countries, and even in whole regions, serious problems already exist and could deepen.

Diets in developing countries are changing as incomes rise. The share of staples, such as cereals, roots and tubers, is declining, while that of meat, dairy products and oil crops is rising. Between 1964-1966 and 1997-1999, per capita meat consumption in developing countries rose by 150% and that of milk and dairy products by 60%. By 2030, per capita consumption of livestock products could rise by a further 44%. Poultry consumption is predicted to grow the fastest. Productivity improvements are likely to be a major source of growth. Milk yields should improve, while breeding and improved management should increase average carcass weights and off-take rates. This will allow increased production with lower growth in animal numbers, and a corresponding slowdown in the growth of environmental damage from grazing and animal wastes.

In developing countries, demand is predicted to grow faster than production, resulting in a growing trade deficit. In meat products this deficit will rise steeply, from 1.2 million tonnes per year in 1997-1999 to 5.9 million tonnes per year in 2030 (despite growing meat exports from Latin America), while in the case of milk and dairy products, the rise will be less steep but still considerable, from 20 million tonnes per year in 1997-1999 to 39 million tonnes per year in 2030. An increasing share of livestock production will probably come from industrial enterprises. In recent years, production from this sector has grown twice as fast as that from more traditional mixed farming systems and more than six times faster than that from grazing systems.

Below is a list of protein content in foods, organised by food group and given in measurements of grams of protein per 100 grams of food portion. Most natural foods are composed largely of water. Reduction of water content has the greatest effect of increasing protein. It is to be noted that not all protein is equally digestible. Protein Digestibility Corrected Amino Acid Score (PDCAAS) is a method of evaluating the protein quality based on the amino acid requirements of humans. In spcic groupings we see:

Cheese

Protein content range: 7.0 to 40.8

- high scores: parmesan 34.99 to 40.79; gruyère 29.8; Edam 25; traditional cheddar 24.9 to 27.2
- average scores: camembert 19.8; processed cheddar 16.42 to 24.6
- low scores: feta 14.7; ricotta 11.26 to 11.39

Milk and milk substitutes

- Cow milk (fluid, raw or pasteurized) - 3.2 to 3.3
- Soy milk 5.1 to 7.5
- Goat milk 4.9 to 9.9

Common red meats

- Beef, cooked - 16.9 to 40.6
    - high scores: braised eye-of-round steak 40.62; broiled t-bone steak (porterhouse) 32.11
    - average scores: baked lean (ground beef) 24.47
    - low scores: corned beef: 16.91
- Lamb, cooked - 20.91 to 50.9

White meat

- Chicken: 31.07

- Mock meat (cooked vegetarian preparations): 18.53 to 23.64

Vegetables

- Nori seaweed, dried sheets: 5.81
- ready-to-eat green vegetables: 0.33 to 3.11
- ready-to-eat starchy tubers: 0.87 to 6.17
- boiled Black Beans: 9
- boiled chia seeds: 16
    - high scores: home-prepared potato pancakes 6.17; French fries 3.18-4.03
    - average scores: baked potato 2.5; boiled yam 1.49
    - low scores: boiled sweet potato 0.87.

Legumes

- dry roasted soybeans: 13
- boiled lentils: 9
- boiled Green Peas: 5
- boiled Black eyed beans: 8
- boiled chickpeas: 9
- peanuts (raw, roasted, butter): 23.68 to 28.04

Baked products
- Bread: 6.7 to 11.4

- Crackers: 7.43

Carbon footprint of production in kg CO2e

Milk and milk substitutes

- Cow milk (fluid, raw or pasteurized) 0.46

Common red meats

- Beef, cooked 8.0
- Lamb, cooked 11.3

White meat

- Fish 2.38
- Chicken 1.7

Vegetables

- ready-to-eat green vegetables: 0.09
- ready-to-eat starchy tubers: 0.15
- boiled Black Beans: 0.22

Legumes

- dry roasted soybeans: 0.25
- boiled lentils: 0.24
- boiled chickpeas: 0.29
- peanuts (raw, roasted,): 0.26
- peanut butter: 0.69

Baked products
- Bread: 0.45

. The world's livestock sector is growing at a break-neck pace and the driving force behind this enormous surge is a combination of population growth, rising incomes, and urbanized age. Meat production is projected to increase from 218 million tonnes in 1999 to 476 million tonnes by 2030, a seventy percent increase.

There is a strong relationship between the level of income and the consumption of animal protein, with the consumption of meat, milk, and eggs increasing at the expense of staple foods. Because of the recent steep declines in prices developing countries are embarking on a higher meat consumption at much lower levels of gross domestic product than the industrialized countries did some 30 years ago.

Urbanization is a major driving force influencing global demand for livestock products. Urbanization sin that stimulates improvements to infrastructure, including cold chains which permit trade in perishable goods. Compared with the less diverse of my diets of rural communities, city toys have are buried to diet rich in animal proteins and fats, and characterized by higher consumptions of meat. There has been a remarkable increase in the consumption of an old products in countries such as Brazil and China, although the levels are still well below the levels of consumption in North America and most other industrialized countries.

As diets become richer and more diverse the high-value protein that livestock sectors offer improves the nutrition of the vast majority of the world livestock products or only provide high-value protein but are also important sources of a wide range of essential micronutrients, in particular minerals such as iron and sink and vitamins such as vitamin a. Four large majorities of persons in the world, particularly in developing countries, livestock products remain a desired or nutritional value and takes excessive consumption of animal products in some countries and social classes can however lead to excessive intakes of flat.

The growing demand for livestock products is likely to have an undesirable impact on the environment. Consideration must be given the fact that the carbon footprint of the need for direct red meat production years of 32 times the ratio equivalent carbon footprint of plant-based proteins. Attempts have been made to estimate the environmental impact of industrial livestock production; for instance it has been estimated that the number of people fled in a year per hectare ranges from a high 22 for potatoes to a low of 1 for beef and lamb. The low energy conversion ratio for feed to need is another concern, since some of the cereal grain food produced must be converted to livestock production. Likewise, land and water requirements for meat production is likely to become a major concern with the increasing demand for animal product resulting in more intensive livestock production systems and distribution.

Fish Farming

Despite fluctuations in supply and demand caused by the changing state of fisheries resources, the economic climate and environmental conditions, fisheries including a culture have traditionally been, and remain an important source of food, employment, and revenue and make countries and communities. After the remarkable increase in both marine and inland capture of British during the period 1950 to 1960 world fisheries production has leveled off since the 70s. It is therefore very unlikely but substantial increases in total catch will be obtained in the future in contrast, aquaculture has followed the opposite path. Starting from insignificant total production, inland and marine aquaculture production has been growing at a remarkable rate, offsetting part of the reduction of the ocean catch of fish.

The protein derived from fish, crustaceans and mollusks account for between 13.8% and 16.5% of the total animal protein intake of the human population. The average apparent per capita consumption increased from 9 kg per year in the early 60s to 16 kg in 1997. World fisheries production has kept ahead of population growth over the past three decades. Total fish production has almost doubled, from 65 million tonnes in

1970 to 125 million tonnes in 1999, when the world average intake of fish, crustaceans and molluscs reached 16.3 kg per person. By 2030, annual fish consumption is likely to rise to some 150-160 million tonnes, or between 19-20 kg per person. This amount is significantly lower than the potential demand, as environmental factors are expected to limit supply. During the 1990s the marine catch levelled out at 80-85 million tonnes per year, and by the turn of the century, three-quarters of ocean fish stocks were overfished, depleted or exploited up to their maximum sustainable yield. Further growth in the marine catch can only be modest. Aquaculture, which currently amounts to just more than thirty percent of market fish will need to expand in order to meet the new targets.

Aquaculture compensated for this marine slowdown, doubling its share of world fish production during the 1990s. It is expected to continue to grow rapidly, at rates of 5-7% per year. In all sectors of fishing it will be essential to pursue forms of management conducive to sustainable exploitation, especially for resources under common ownership or no ownership. Aquaculture, also known as aquafarming, is the farming of fish, crustaceans, molluscs, aquatic plants, algae, and other aquatic organisms. Aquaculture involves cultivating freshwater and saltwater populations under controlled conditions, and can be contrasted with commercial fishing, which is the harvesting of wild fish. Mariculture refers to aquaculture practiced in marine environments and in underwater habitats.

According to the FAO, aquaculture "Farming implies some form of intervention in the rearing process to enhance production, such as regular stocking, feeding, protection from predators, etc. Farming also implies individual or corporate ownership of the stock being cultivated." The reported output from global aquaculture operations in 2014 supplied over one half of the fish and shellfish that is directly consumed by humans; however, there are issues about the reliability of the reported figures. Further, in current aquaculture practice, products from several pounds of wild fish are used to produce one pound of a piscivorous fish like salmon.

Particular kinds of aquaculture include fish farming, shrimp farming, oyster farming, mariculture, algaculture (such as seaweed farming), and the cultivation of ornamental fish. Particular methods include aquaponics and integrated multi-trophic aquaculture, both of which integrate fish farming and plant farming.

Harvest stagnation in wild fisheries and overexploitation of popular marine species, combined with a growing demand for high-quality protein, encouraged aquaculturists to domesticate other marine species.\ At the outset of modern aquaculture, many were optimistic that a "Blue Revolution" could take place in aquaculture, just as the Green Revolution of the 20th century had revolutionized agriculture. Although land animals had long been domesticated, most seafood species were still caught from the wild. Concerned about the impact of growing demand for seafood on the world's oceans, prominent ocean explorer Jacques Cousteau wrote in 1973: "With earth's burgeoning human populations to feed, we must turn to the sea with new understanding and new technology."\

About 430 (97%) of the species cultured as of 2007 were domesticated during the 20th and 21st centuries, of which an estimated 106 came in the decade to 2007. Given the long-term importance of agriculture, to date, only 0.08% of known land plant species and 0.0002% of known land animal species have been domesticated, compared with 0.17% of known marine plant species and 0.13% of known marine animal species. Domestication typically involves about a decade of scientific research. Domesticating aquatic species involves fewer risks to humans than do land animals, which took a large toll in human lives. Most major human diseases originated in domesticated animals, including diseases such as smallpox and diphtheria, that like most infectious diseases, move to humans from animals. No human pathogens of comparable virulence have yet emerged from marine species.

Biological control methods to manage parasites are already being used, such as cleaner fish (e.g. lumpsuckers and wrasse) to control sea lice populations in salmon farming. Models are being used to help with spatial planning and siting of fish farms in order to minimize impact.

The decline in wild fish stocks has increased the demand for farmed fish. However, finding alternative sources of protein and oil for fish feed is necessary so the aquaculture industry can grow sustainably; otherwise, it represents a great risk for the over-exploitation of forage fish.

Many new natural compounds are discovered every year, but producing them on a large enough scale for commercial purposes is almost impossible.

It is highly probable that future developments in this field will rely on microorganisms, but greater funding and further research is needed to overcome the lack of knowledge in this field

Aquaculture is an especially important economic activity in China. Between 1980 and 1997, the Chinese Bureau of Fisheries reports, aquaculture harvests grew at an annual rate of 16.7%, jumping from 1.9 million tonnes to nearly 23 million tonnes. In 2005, China accounted for 70% of world production. Aquaculture is also currently one of the fastest-growing areas of food production in the U.S..

About 90% of all U.S. shrimp consumption is farmed and imported. In recent years, salmon aquaculture has become a major export of southern Chile, especially in Puerto Montt, Chile's fastest-growing city.

A United Nations report titled *The State of the World Fisheries and Aquaculture* released in May 2014 maintained fisheries and aquaculture support the livelihoods of some 60 million people in Asia and Africa

Chapter 8 Genetically Modified Organisms

The 4th Green Revolution is generally reserved for Genetically Modification of Agricultural products. For many people in the First World, genetically modified crops have become the latest incarnation of evil: biotechnology, which sacrifices humans and the environment for the sake of revenues and shareholder value. On one side of the heated discourse are people who firmly believe that GM crops pose a threat to human health and biodiversity. On the other side are philosophers and scientists who are convinced that genetic engineering of plants represents a technology with enormous potential for increasing food production in an environmentally benign way. This controversy has to some extent degenerated into a sterile, even hysterical debate, where important facts are largely ignored and where relatively few new ideas are introduced in order to find ways for using this technology in the safest possible manner.

The opposition to GM crops [including animal stock] is in part due to the fact that most consumers in the First World have not yet seen any direct advantages of products derived from this new technology, be it lower prices or improved nutritional quality. Given the apparent lack of benefit, many consumer associations and environmental groups think it is unjustified to accept any possible risk to the environment that might come from the use of GM crops.

Many critics trust neither industry nor regulatory agencies, which they regard as allies of the chemical industry and biotechnology companies. The propaganda from some non-governmental groups, usually exerted through irresponsible journalism, has led to a serious deterioration of public confidence in scientists and governmental regulation institutions. Destruction of test sites by the most radical environmentalist groups, proposed moratoria on transgenic crops and food retailers refusing to sell transgenic food products are just some of the manifestations that have sprung from the adamant opposition against GM technology. Unfortunately, this has happened without an open, sensible and serious discussion of the scientific, economic and political facts.

Most agricultural scientists would consider transgenic crops as safe as, or even safer, for the environment than comparable products obtained through traditional breeding. However, some scientific journals have published negative reports about the safety of GM crops, such as the potentially harmful effects of pollen from insect-resistant corn on the larvae of the monarch butterfly. This publication, as well as its exaggeration and manipulation by environmentalists, has increased the public pressure on the regulatory authorities of various countries to prohibit or delay the use of GM crops.

But while environmental and consumer advocates in the First World fight against the worldwide use of GM crops in agriculture, hundreds of millions of people in the Third World are malnourished. And while trying to protect the environment and consumers in developed countries, critics of GM crops block a technology that could be of immense benefit for the majority of people in the Southern Hemisphere. Any serious attempt to discuss and make long-term decisions regarding GM plants must therefore take into account the facts about poor countries that, so far, have been largely ignored by opponents of this technology.

The human population is growing and it is growing faster than anticipated. In 2016, the UN published its latest estimates, which project the world's population to be 9.3 billion in 2050—400 million more than previously estimated. To feed all of these people and thus prevent famine, upheaval or civil war, more and better food is needed, at least for the majority of people on this planet who need it most. Opponents of GM crops claim that feeding the poor is only a matter of better distribution. But inadequate distribution occurs even in developing countries that are net exporters of agricultural products. Thus, to ensure that food is available to everybody, local food production in poor countries must increase. This will also benefit the economies of these countries and reduce their dependence on the industrialised world.

Farmers in general are neither in favour of, nor against GM crops. They adopt whatever technologies promise them lower production costs, increased productivity or products of higher value. Indeed, GM crops have been used not only in the USA but also in Argentina, China and Mexico, showing that farmers in developing countries benefit from their cultivation.

The potential benefits and risks of a new technology should be assessed only by comparing it to the technology it will replace. It is futile to judge a technology in isolation. Another important point often left out of the debate is how to make sure that new technologies help people in developing countries. Some people argue that GM technology is controlled by large multinational companies and thus will never be used by small farmers. Consequently, instead of condemning and blocking GM crop technology, government-funded institutions and non-governmental organisations should find ways to ensure that the knowledge is transferred to developing countries.

Non-governmental organisations insist that the voice of the public at large, as opposed to only scientists, should be heard and taken into account. Certainly, everybody agrees with this position. However, one wonders which public these organisations refer to. Do they represent public opinion in developing countries? Do they really know the problems and needs of small farmers in developing countries?

Many readers assume that GM technology is meant to replace traditional breeding and that it will solve all current agricultural problems. It is important to understand that solving the problem of food production for a growing population without harming the environment will require the concerted use of traditional breeding and organic farming, as well as GM crop technology, each being used to solve specific problems and needs. Alleviation of hunger cannot depend on a single technology.

Over the next 50 years, humankind's greatest challenge will be to ensure sufficient food production on a global scale. This means eventually increasing agricultural productivity in tropical areas where crop yields are significantly lower than in temperate climate zones. Here, losses due to pests, plant diseases and poor soils are exacerbated by climatic conditions that favour the proliferation of insect pests and disease vectors. In addition, post-harvest losses in tropical areas are higher than elsewhere due to fungal and insect infestations, as well as the lack of appropriate storage facilities. Despite efforts to prevent such crop losses, pests destroy more than half of the world wide food production. Insect damage, the majority of which occurs in the developing world, is responsible for around 15% of the world's pre-harvest food losses. Future food production will be further impaired by the global marketplace as developed countries eliminate subsidies for the production of basic staples such as cereals, meat and dairy products.

Doing nothing to help feed a growing population also puts the environment at risk. Tropical forests are irreplaceable regional and global ecosystems that contain more than 90% of plant and animal species. But more than 11 million hectares of forest are cleared every year by farmers searching for more productive land. Indiscriminate conversion of tropical forest into agricultural land will have far greater ecological impact than the use of GM crops or any other technology (Fedoroff and Cohen, 1999).

Without having access to GM technology, the only alternative for Third World countries to increase food production would be to use more fertilisers, insecticides and herbicides–certainly not beneficial to the environment either. Furthermore, most farmers in poor countries simply cannot afford these chemicals that have been developed for large mechanised farms in the First World. GM technology has already demonstrated that it has the potential to increase food production while decreasing production costs. For virus-, insect- and herbicide-resistant plants, an average increase in yield of 5–10%, up to 40% saved on herbicides and savings of US$ 60 to 120 per acre on insecticides have been reported.

But such resistant plants, despite their impressive economic and environmental value, will have only a limited impact on global food production. Most GM crops currently available on the market were developed with the aim of reducing production costs in agricultural areas that already have high productivity levels, or of increasing the final value of the product, for instance, by improving oil quality. So far, plant gene transfer technology and research on transgenic plant varieties have been driven by the potential market value of the desired trait, determined by farmers in the USA and Western

Europe. Because biotechnology companies have to make considerable investments to develop, test and commercialise transgenic plants, this is understandable. But in terms of global food production, it will be necessary to ensure that GM technology is made available to developing countries where researchers can create or vary crops adapted to local conditions. This would also help to facilitate the survival of small farms and their cultural traditions in these countries.

Agriculture in tropical and subtropical regions faces specific problems that are different from those that limit food production in the First World. Since many of these problems are common to many countries and affect a wide spectrum of crops, potential solutions that can be applied to different plant species are urgently needed. Unfortunately, this is not considered as a research priority in developed countries, and little is being done to address these problems. It is also unfortunate that most developing countries do not have the resources to invest in the biotechnology needed to increase their agricultural productivity within the time frame required to cope with the increasing demand for food.

Even when there is a clear benefit arising from GM technology in a poor country, its application is often vulnerable to opposition from advocates of environmental or consumer groups. A possible solution would be the medium- and long-term monitoring of transgenic releases to investigate potential harm to the environment. This is already being implemented or planned by the USA and Japan (Reichhardt, 1999; Saegusa, 1999). The scientific community is already supporting the view that release of GM plants should be approved only if there is negligible or very low risk, and that such a finding may still be subject to confirmation or modification through the collection and analysis of field data. On the other hand, NGOs and the public should accept a decision by the regulating authorities to release GM crops after extensive monitoring has shown no damage to, or negative influence on the environment. Furthermore, monitoring would also help to detect potential environmental harm early in the process, and thus allow authorities to ban traits that are responsible for the harm.

Many thousands of people still starve to death and nearly 800 million are malnourished (Fedoroff and Cohen, 1999). If only a few major companies control the technology [and small farmers in poor countries do not fall into the category] problems loom. Fortunately, multinational companies have shown an increased interest in donating technology to developing countries, and some technology transfer is already under way (Qaim, 1998).

To ensure safe and sufficient food production, political and economic decisions by governments and companies, rather than technological limitations, will determine how successfully we can feed a growing population in poor countries. In order to make wise decisions, an international body should be created to ensure that the necessary technology reaches the places where it is needed and to deal with the political, economic and social problems associated with technology transfer. UNESCO has been designating monuments as belonging to humankind, which must be preserved not only for the benefit of the locals, but for the entire world. Perhaps new technologies that could solve fundamental problems of human well-being should be given a similar status, to ensure that they reach everybody who needs them.

Pest infestations, diseases and poor weather conditions may all significantly lower crop yields in developing countries. GM crops could address these problems, where other breeding techniques have failed. We consider a series of case studies:

Half the cotton grown in China in 2002 was genetically modified to produce a substance that is poisonous to the cotton bollworm, a pest that devastates many cotton crops. Farmers had previously applied the toxin directly by spraying the crops. The benefits of the 'Bt cotton' are a reduction in pesticide use, an increase in yields and profits, and health benefits for farm workers who often apply pesticides without protective clothing.

Plants can be genetically modified to be resistant to bacterial, fungal or viral infestation. Examples include research on sweet potatoes to improve viral resistance and bananas modified to resist the Black Sigatoka fungus. Untreated, this fungus can reduce banana yields by as much as 70% but fungicides are expensive.

Resistance to drought, heat, frost, acid or salty soil may be sought in differing environmental exposures. A gene from a plant which can survive prolonged water stress in desert conditions has been introduced into rice. This allows rice to produce a sugar that protects the plant during dehydration, allowing it to survive periods of drought.

Plants can be genetically modified to be tolerant to a specific weedkiller. This allows farmers to control a wide range of weeds with less weedkiller while not affecting the modified crop. Herbicide tolerant crops are grown mainly in developed countries. However, recently they have been used in some developing countries. For example, more than 90% of soybeans grown in Argentina during 2002 were GM.

Crops can be genetically modified to contain additional nutrients that are lacking from the diets of many people in developing countries. One example is Golden Rice, which has been modified to have enhanced levels of ß-carotene, in order to help to prevent vitamin A deficiency. 14 million children under five suffer clinically from this deficiency, which can cause childhood blindness.

Banana wilt was first seen in Uganda in 2001, and neither pesticides nor chemicals have stopped it. Farmers tried to control the wilt's spread by torching infected plants and disinfecting tools, but the disease cut Ugandan banana yields by as much as half from 2001 to 2004. In the country's central region, wilt hit 80 percent of plants, and sometimes knocked out whole fields, according to a report from *The Guardian*. Scientists at Uganda's National Agricultural Research Organization (NARO) — which receives funding from the Gates Foundation — created a genetically modified banana by inserting a green pepper gene into the banana's genome. The new gene seems to trigger a process that kills infected cells and saves the plant. NARO wants to give the seeds away for free, but no regulation exists around GMOs in Uganda, and Uganda is obligated to take a cautionary approach to GMO technology, as signer of 2000's Cartagena protocol. The Ugandan government is considering passing a law that would allow the introduction of GMOs, including the bacteria-resistant banana, but some food scientists worry it may open the door to corporate exploitation by multinational companies like Monsanto down the line.

Farming.and Poverty

This year, the Gates Foundation's annual letter points to innovations in farming as a revolution that will transform the lives of the poor over the next 15 years, particularly in Africa. Food is a fundamental human right; nonetheless, people are starving. The UN's World Food Programme estimates over 800 million individuals, or one in nine people on the planet, struggle to find enough food to eat on a regular basis. In places like Sub-Saharan Africa, hunger is a tremendous problem — and an ironic one. The region is home to abundant arable land; 70 percent of the population there farms. But the prevalence of hunger there is also the highest in the world — one in five people are undernourished. Chronic malnutrition has stunted the growth of 40 percent of children under the age of five, according to UNICEF. That is 25 million children.

In Sub-Saharan Africa, hunger is a tremendous problem — and an ironic one. A new generation of highly productive crops, Gates suggests, are part of the solution to address global hunger — seeds that are drought-resistant, disease-resistant, productive, and nutritious could benefit farmers. Some of the crops can be bred through traditional methods, but Gates thinks many African countries will adopt GMOs, or genetically modified organisms. GMOs are an accelerated version of the traditional methods of plant breeding which require raising several generations of plants, improving their yield or drought-tolerance properties over years if not decades. But genetic information lets scientists tweak specific genes — a much faster process. It also expands the range of possible alterations, since genes from one species can be inserted into another.

North American Experience

The first American GMO crop was the Flavr Savr tomato, created by California company Calgene and green-lit by the US Food and Drug Administration in 1994. The modified tomatoes didn't get squishy as quickly as regular tomatoes. Though Flavr Savr tomatoes are no longer sold — turns out, it's more economical to ship green tomatoes — genetic modifications caught on and proved more successful in other foods. More than 90 percent of soybeans and 80 percent of corn sold in the US today are GMOs created by companies like Monsanto, Cargill, and Dupont. GMO seeds are often more expensive than conventionally bred varieties, a concern voiced by some opponents of Uganda's bill.

More than 90 percent of soybeans and 80 percent of corn sold in the US today are GMOs. GMOs have been widely touted as a solution to hunger and malnutrition: engineering for specific traits, like the wilt-resistant banana in Uganda, they could make farmers less vulnerable to crop loss. "GMO-derived seeds will provide far better productivity, better drought tolerance, salinity tolerance, and if the safety is proven, then the African countries will be among the biggest beneficiaries," Bill Gates posited.

The Gates Foundation Asset Trust, which manages the foundation's assets, has previously held shares of Monsanto. The trust hasn't held shares of Monsanto "for a few years" says Alex Reid, a foundation spokesperson, who adds that the trust is managed separately and doesn't get input into what the foundation funds.

The Gates Foundation suggests that by using better fertilizer and more productive crops such as GMOs, African farmers could "theoretically double their yields." (The average yield per acre in Africa is one-fifth of that in the US.) "With the right investments," the Gates letter goes on, it may be possible for farmers on the continent to "increase productivity by 50 percent overall." The Foundation believes so strongly in the promise of farming that in 2013 it spent 22 percent of its $1.8 billion global development expenditure on agriculture. But even if GMO crops yield more produce, will that translate to less hunger?

Lowered production is an issue, certainly. Once, Africa was a major food exporter, sending out coffee, cocoa, and spices — but the price of commodities dropped in the 1980s, and imports have outpaced exports. Food production has mostly been stagnant since then, while consumption has grown, according to the UN's Food and Agriculture Organization. Today, African countries spend $35 billion to $40 billion a year on imported food. Relying heavily on imports raises the price of regionally produced food, contributing to a cycle of poverty.

Increasing production, however, may be a red herring: according to the World Food Programme, the planet actually produces enough food to feed everyone alive more than 2,000 calories a day. But global funding priorities remain "heavily focused on increasing agricultural production," according to a 2013 report from the United Nations Conference on Trade and Development. "The perception that there is a supply-side problem is, however, questionable," the report reads. "Hunger and malnutrition are mainly related to a lack of purchasing power and/or the inability of the rural poor to be self-sufficient. Meeting the food security challenges is primarily about the empowerment of the poor and their food sovereignty." In other words: the primary problem isn't technology, it's distribution.

Feeding a Hungry Planet

The planet already produces enough food to feed everyone alive more than 2,000 calories a day. Poverty exacerbates the issue: in some developing countries, farmers can't afford the seeds they plant to feed their families; some are too poor to even feed themselves well, limiting their ability to work, earn more money, or invest in their own farms. "In a lot of famines, you have food leaving the famine areas, because buying power has collapsed," says Gawain Kripke, the director of policy and research at Oxfam America. (Oxfam America has previously received funding from the Gates Foundation.) "It's not supply or availability of food. There's just no buying power."

Farmers in Africa are also politically weak, says Kripke. As a result, countries' policies favor urban consumers in budget allocations. That makes it difficult for farmers to lobby for the changes they'd need to create a functioning agricultural supply chain. The stance is supported by the UN's FAO, which says, in a 2014 report, that the keys to progress in nutrition include "sustained political commitment to food security at the highest level," as well as policies that give a voice to politically weak groups. The Gates Foundation donated $12 million to the FAO last year.

About 30 percent of crops produced in Africa are lost after harvest. Even assuming the technology makes it to the needy, farmers don't necessarily have a reason to grow more food than they can eat themselves. Without basic infrastructure like markets, storage, and trade agreements, there's little they can do with the extra produce. Of the 54 states within the African Union, only 10 have allocated at least a tenth of their public investments to agriculture — that is, to infrastructure, irrigation, research, and development. Without markets, international trade, places to store surplus harvests, and irrigation, it's hard to encourage farmers to invest in growing more than they need.

Before heading up AGRA, Kalibata was the minister of Rwanda's Agriculture and Animal Resources. Between 2004 and 2014, Rwanda more than doubled its crop production by increasing agriculture investment from 3.5 percent to 7.2 percent of government expenditures. That investment reduced poverty by almost a third, according to AGRA's statistics.

Using the lessons Kalibata learned in Rwanda, AGRA is trying to promote policy changes, including increased government investment, across the continent. "One policy breakthrough in one country could be a big breakthrough for everyone," she says. In 2013, the Gates Foundation awarded the group $4 million for policy change and about $23 million for education and technology.

AGRA advocates for basic, but wide-reaching structural changes: for instance, Africa has no continent-wide regulations on grades, standards, weights, or measures — all of which complicate trade. There are also no strong regulations on how food should be stored in warehouses, which may mean losses for farmers if the storehouses aren't up to grade. Customs procedures are often cumbersome, and vary widely — which drive up the cost of transactions.

"All the infrastructure we take for granted in a modern agricultural supply chain doesn't exist [in Africa]," says Oxfam America's Kripke. "So where do you start? Seeds? Roads?"

Africa has no continent-wide regulations on grades, standards, weights, or measures — all of which complicate trade. Roads are a safe bet, according to International Food Policy Research Institute. In fact, they're one of the best investments governments can make. Not only do they get seeds and other agricultural products and technology into an area, roads are crucial for trade. Without access to roads, farmers tend to buy seeds at high prices and sell produce at lower ones, which exposes them to more risk from food price fluctuations. Sometimes, farmers are unable to sell their crops at all. And when that happens, farmers become apathetic about adopting new technology: with little incentive of increasing yield if most of it will sit and spoil

Infrastructure investment has a proven impact: the introduction of roads in 15 rural villages in Ethiopia increased consumption by households by about 16 percent, while lowering poverty about 7 percent, according to a 2008 IFPRI study.

More roads may also mean more information: currently agricultural extension programmes — which help train farmers and educate them about the marketplace — can be tough for farmers to reach. That's particularly true for about half of African farmers: women, who are less able to travel because of familial responsibilities or the hazards of the road. "Gender is a pretty big concern," says Kripke. "Especially women farmers have trouble," AGRA's Kalibata says. And farm education is an area where women especially stand to gain, says Calestous Juma, a professor at Harvard's Kennedy School of Government and director of the Science, Technology, and Globalization Project at the Belfer Center for Science and International Affairs. Juma also serves as the director of Agricultural Innovation in the Africa Project, which the Gates Foundation funds.

The Uganda Experience

All these factors come into play in Uganda. In this sub-Saharan nation of 37 million, bananas are essential — about 75 percent of farmers grow them there. The country depends on agriculture for 23 percent of its gross domestic product. But like other countries in the region, Uganda suffers from a dearth of infrastructure. Though the country is only slightly smaller than the UK, it has only 2,028 miles (3,264 km) of paved roads, compared to the UK's 245,068 miles (394,428 km). That makes trade and transit quite difficult, especially since 85 percent of Uganda's 36 million people live in rural areas. A quarter of the population is below the poverty line. As a result, almost 40 percent of children there are undernourished, according to the US government's Feed the Future initiative.

Can the genetically modified bananas have a positive impact for millions of Ugandans? "Many farmers can't afford expensive seeds. They would have no rights." Uganda's wilt-resistant banana is the best possible scenario for GMO adoption, in some respects. The strain was created by local scientists and, because it's being distributed for free, won't lead to capital from farmers flowing out of the country. But some activists are concerned that the banana GMO will open the door to other crops with pernicious consequences. "Farmers have been told that the GMOs are almost the same as non-GMOs," Ellady Muyambi, an environmental scientist at the Uganda Network on Toxic Free Malaria, told NPR. "But they would have to go to a company to buy the seeds. Many farmers can't afford expensive seeds. They would have no rights."

Hunger in Uganda is a bigger issue than the impact of one GMO law: evidence suggests that improving farmer education programmes and infrastructure investment can have a bigger impact than increasing productivity alone. Without infrastructure — roads, granaries, markets, irrigation — and policy changes, more food won't eliminate hunger. And while GMOs like the wilt-resistant banana may save critical crops, it's not clear they can ensure food security in hungry communities.

Uganda has bigger problems on the horizon. According to the Brookings Institute, Uganda should expect inflation in 2015 — the result of government borrowing, a depletion of foreign currency reserves, and cutbacks on essential supplies — that will "widen the income gap" and "reverse the gains made in poverty reduction." In other words: the economic situation in Uganda is set to reduce the purchasing power of the impoverished. If the Brookings Institute is right, that means hunger, with or without genetically modified bananas.

Chapter 9 GMO Growth

Farmers have widely adopted GM technology. Between 1996 and 2015, the total surface area of land cultivated with GM crops increased by a factor of 100, from 17,000 km² (4.2 million acres) to 1,797,000 km² (444 million acres). 10% of the world's arable land was planted with GM crops in 2010. In the US, by 2014, 94% of the planted area of soybeans, 96% of cotton and 93% of corn were genetically modified varieties. Use of GM crops expanded rapidly in developing countries, with about 18 million farmers growing 54% of worldwide GM crops by 2013. A 2014 meta-analysis concluded that GM technology adoption had reduced chemical pesticide use by 37%, increased crop yields by 22%, and increased farmer profits by 68%. This reduction in pesticide use has been ecologically beneficial, but benefits may be reduced by overuse. Yield gains and pesticide reductions are larger for insect-resistant crops than for herbicide-tolerant crops. Yield and profit gains are higher in developing countries than in developed countries.

There is a scientific consensus that currently available non-foods derived from GM crops poses no risk to human. Nonetheless, members of the public are much less likely than scientists to perceive GM 'products' as safe. The legal and regulatory status of GMOs varies by country, with some nations banning or restricting them by applying food standards to products destined for other use (viz. Corn in Iowa totally pre-sold for biofuel), and others permitting them with widely differing degrees of regulation.

Opponents of GMOs have objected to GM crops on several grounds, including environmental concerns, whether food produced from GM crops is safe, whether GM crops are needed to address the world's food needs, and concerns raised by the fact these organisms are subject to intellectual property law.

DNA transfers naturally between organisms. Several natural mechanisms allow gene flow across species. These occur in nature on a large scale – for example, it is one mechanism for the development of antibiotic resistance in bacteria. This is facilitated by transposons, retrotransposons, proviruses and other mobile genetic elements that naturally translocate DNA to new loci in a genome. Movement occurs over an evolutionary time scale.

The introduction of foreign germplasm into crops has been achieved by traditional crop breeders by overcoming species barriers. A hybrid cereal grain was created in 1875, by crossing wheat and rye. Since then important traits including dwarfing genes and rust resistance have been introduced. Plant tissue culture and deliberate mutations have enabled humans to alter the makeup of plant genomes.

The first genetically modified crop plant was produced in 1982, an antibiotic-resistant non-food tobacco plant. The first field trials occurred in France and the USA in 1986, when tobacco plants were engineered for herbicide resistance. In 1987, Plant Genetic Systems (Ghent, Belgium), founded by Marc Van Montagu and Jeff Schell, was the first company to genetically engineer insect-resistant (tobacco) plants by incorporating genes that produced insecticidal proteins from *Bacillus thuringiensis* (Bt).

The People's Republic of China was the first country to allow commercialized transgenic plants, introducing a virus-resistant tobacco in 1992, which was withdrawn in 1997. In 1994, the European Union approved tobacco engineered to be resistant to the herbicide bromoxynil, making it the first commercially genetically engineered crop marketed in Europe.

In 1995 canola with modified oil composition (Calgene), Bt maize (Ciba-Geigy), bromoxynil-resistant cotton (Calgene), Bt cotton

(Monsanto), glyphosate-resistant soybeans (Monsanto), all non-food destined patents, were approved. As of mid-1996, a total of 35 approvals had been granted to commercially grow 8 transgenic crops and one flower crop (carnation), with 8 different traits in 6 countries plus the EU. Genetically engineered crops have genes added or removed using genetic engineering techniques, originally including gene guns, electroporation, microinjection and agrobacterium. More recently, CRISPR and TALEN offered much more precise and convenient editing techniques.

Gene guns (also known as biolistics) "shoot" (direct high energy particles or radiations against target genes into plant cells. It is the most common method. DNA is bound to tiny particles of gold or tungsten which are subsequently shot into plant tissue or single plant cells under high pressure. The accelerated particles penetrate both the cell wall and membranes. The DNA separates from the metal and is integrated into plant DNA inside the nucleus. This method has been applied successfully for many cultivated crops, especially monocots like wheat or maize. The major disadvantage of this procedure is that serious damage can be done to the cellular tissue.

Agrobacterium tumefaciens-mediated transformation is another common technique. Agrobacteria are natural plant parasites, and their natural ability to transfer genes provides another engineering method. To create a suitable environment for themselves, these Agrobacteria insert their genes into plant hosts, resulting in a proliferation of modified plant cells near the soil level (crown gall). The genetic information for tumour growth is encoded on a mobile, circular DNA fragment (plasmid). When *Agrobacterium* infects a plant, it transfers this T-DNA to a random site in the plant genome. When used in genetic engineering the bacterial T-DNA is removed from the bacterial plasmid and replaced with the desired foreign gene. The bacterium is a vector, enabling transportation of foreign genes into plants. This method works especially well for dicotyledonous plants like potatoes, tomatoes, and tobacco. Agrobacteria infection is less successful in crops like wheat and maize.

Electroporation is used when the plant tissue does not contain cell walls. In this technique, "DNA enters the plant cells through miniature pores which are temporarily caused by electric pulses." Microinjection directly injects the gene into the DNA.

Plant scientists, backed by results of modern comprehensive profiling of crop composition, point out that crops modified using GM techniques are less likely to have unintended changes than are conventionally bred crops.

In research tobacco and *Arabidopsis thaliana* are the most frequently modified plants, due to well-developed transformation methods, easy propagation and well studied genomes. They serve as model organisms for other plant species.

Introducing new genes into plants requires a promoter specific to the area where the gene is to be expressed. For instance, to express a gene only in rice grains and not in leaves, an endosperm-specific promoter is used. The codons of the gene must be optimized for the organism due to codon usage bias.

Transgenic plants have genes inserted into them that are derived from another species. The inserted genes can come from species within the same kingdom (plant to plant) or between kingdoms (for example, bacteria to plant). In many cases the inserted DNA has to be modified slightly in order to correctly and efficiently express in the host organism. Transgenic plants are used to express proteins like the cry toxins from *B. thuringiensis*, herbicide resistant genes, antibodies and antigens for vaccinations. A study led by the European Food Safety Authority (EFSA) found also viral genes in transgenic plants. Transgenic carrots have been used to produce the drug Taliglucerase alfa which is used to treat Gaucher's disease. In the laboratory, transgenic plants have been modified to increase photosynthesis (currently about 2% in most plants versus the theoretic potential of 9–10%). This is possible by changing the rubisco enzyme (i.e. changing C3 plants into C4 plants), by placing the rubisco in a carboxysome, by adding CO_2 pumps in the cell wall, by changing the leaf form/size.

Plants have been engineered to exhibit bioluminescence that may become a sustainable alternative to electric lighting.

Cisgenic plants are made using genes found within the same species or a closely related one, where conventional plant breeding can occur. Some breeders and scientists argue that cisgenic modification is useful for plants that are difficult to crossbreed by conventional means (such as potatoes), and that plants in the cisgenic category should not require the same regulatory scrutiny as transgenics.

GM economic value to farmers is one of its major benefits, including in developing nations. A 2010 study found that Bt corn provided economic benefits of $6.9 billion over the previous 14 years in five Midwestern US states, more than 98% non-food committed.. The majority ($4.3 billion) accrued to farmers producing non-Bt corn. This was attributed to European corn borer populations reduced by exposure to Bt corn, leaving fewer to attack conventional corn nearby. Agriculture economists calculated that "world surplus [increased by] $240.3 million for 1996. Of this total, the largest share (59%) went to U.S. farmers. Seed company Monsanto received the next largest share (21%), followed by US consumers (9%), the rest of the world (6%), and the germplasm supplier, Delta & Pine Land Company of Mississippi (5%)."

According to the International Service for the Acquisition of Agri-biotech Applications (ISAAA), in 2014 approximately 18 million farmers grew biotech crops in 28 countries; about 94% of the farmers were resource-poor in developing countries. 53% of the global biotech crop area of 181.5 million hectares was grown in 20 developing countries. PG Economics comprehensive 2012 study concluded that GM crops increased farm incomes worldwide by $14 billion in 2010, with over half this total going to farmers in developing countries.

Critics challenged the claimed benefits to farmers over the prevalence of biased observers and by the absence of randomized controlled trials. The main Bt crop grown by small farmers in developing countries is cotton, another non-food crop. A 2006 review of Bt cotton findings by agricultural economists concluded, "the overall balance sheet, though promising, is mixed.

Economic returns are highly variable over years, farm type, and geographical location".

In 2013 the European Academies Science Advisory Council (EASAC) asked the EU to allow the development of agricultural GM technologies to enable more sustainable agriculture, by employing fewer land, water and nutrient resources. EASAC also criticizes the EU's "time-consuming and expensive regulatory framework" and said that the EU had fallen behind in the adoption of GM technologies.

Participants in agriculture business markets include seed companies, agrochemical companies, distributors, farmers, grain elevators and universities that develop new crops/traits and whose agricultural extensions advise farmers on best practices. According to a 2012 review based on data from the late 1990s and early 2000s, much of the GM crop grown each year is used for livestock feed and increased demand for meat leads to increased demand for GM feedcrops. Feed grain is a non-food (human consumptive application) usage as a percentage of total crop production is 70% for corn and more than 90% of oil seed meals such as soybeans. About 65 million metric tons of GM corn grains and about 70 million metric tons of soybean meals derived from GM soybean currently become feed.

In 2014 the global value of biotech seed was US$15.7 billion; US$11.3 billion (72%) was in industrial countries and US$4.4 billion (28%) was in the developing countries. In 2009, Monsanto had $7.3 billion in sales of seeds and from licensing its technology; DuPont, through its Pioneer subsidiary, was the next biggest company in that market. As of 2009, the overall Roundup line of products including the GM seeds represented about 50% of Monsanto's business.

Some patents on GM traits have expired, allowing the legal development of generic strains that include these traits. For example, generic glyphosate-tolerant GM soybean is now available. Another impact is that traits developed by one vendor can be added to another vendor's proprietary strains, potentially increasing product choice and competition. The patent on the first type of *Roundup Ready* crop that Monsanto produced (soybeans) had already expired in 2014, and the first harvest of off-patent soybeans occurs in the spring of 2015. Monsanto has broadly licensed the patent to other seed companies that include the glyphosate resistance trait in their seed products.

In 2014, the largest review yet concluded that GM crops' effects on farming were positive. The meta-analysis considered all published English-language examinations of the agronomic and economic impacts between 1995 and March 2014 for three major GM crops: soybean, maize, and cotton. The study found that herbicide-tolerant crops have lower production costs, while for insect-resistant crops the reduced pesticide use was offset by higher seed prices, leaving overall production costs about the same.

Yields increased 9% for herbicide tolerance and 25% for insect resistant varieties. Farmers who adopted GM crops made 69% higher profits than those who did not. The review found that GM crops help farmers in developing countries, increasing yields by 14 percentage points. The researchers considered some studies that were not peer-reviewed, and a few that did not report sample sizes. They attempted to correct for publication bias, by considering sources beyond academic journals. The large data set allowed the study to control for potentially confounding variables such as fertiliser use. Separately, they concluded that the funding source did not influence study results.

Bio-remediation

Crops grown today, or under development, have been modified with various traits. These traits include improved shelf life, disease resistance, stress resistance, herbicide resistance, pest resistance, production of useful goods such as biofuel or drugs, and ability to absorb toxins and for use in bioremediation of pollution.

Bioremediation is a waste management technique that involves the use of organisms to remove or neutralize pollutants from a contaminated site. According to the United States EPA, bioremediation is a "treatment that uses naturally occurring organisms to break down hazardous substances into less toxic or non toxic substances". Technologies can be generally classified as *in situ* or *ex situ*. *In situ* bioremediation involves treating the contaminated material at the site, while *ex situ* involves the removal of the contaminated material to be treated elsewhere. Some examples of bioremediation related technologies are

- phytoremediation,
- bioventing,
- bioleaching,
- landfarming,
- bioreactor,
- composting,
- bioaugmentation,
- rhizofiltration, and
- biostimulation.

Bioremediation may occur on its own (natural attenuation or intrinsic bioremediation) or may only effectively occur through the addition of fertilizers, oxygen, etc., that help in enhancing the growth of the pollution-eating microbes within the medium (biostimulation). For example, the US Army Corps of Engineers demonstrated that windrowing and aeration of petroleum-contaminated soils enhanced bioremediation using the technique of landfarming. Depleted soil nitrogen status may encourage biodegradation of some nitrogenous organic chemicals, and soil materials with a high capacity to adsorb pollutants may slow down biodegradation owing to limited bioavailability of the chemicals to microbes. Recent advancements have also proven successful via the addition of matched microbe strains to the medium to enhance the resident microbe population's ability to break down contaminants. Microorganisms used to perform the function of bioremediation are known as bioremediators.

Not all contaminants are easily treated by bioremediation using microorganisms. For example, heavy metals such as cadmium and lead are not readily absorbed or captured by microorganisms. Transgenic plants are able to bioaccumulate toxins in their above-ground parts, which are then harvested for removal, mining or wasting.

The heavy metals in the harvested biomass may be further concentrated by incineration or even recycled for industrial use. Some damaged artifacts at museums contain microbes which could be specified as bio remediating agents. In contrast to this situation, other contaminants, such as aromatic hydrocarbons as are common in petroleum, are relatively simple targets for microbial degradation, and some soils may even have some capacity to autoremediate, as it were, owing to the presence or introduction of autochthonous microbial communities capable of degrading these compounds.

The elimination of a wide range of pollutants and wastes from the environment requires increasing our understanding of the relative importance of different pathways and regulatory networks to carbon flux in particular environments and for particular compounds, and they will certainly accelerate the development of bioremediation technologies and biotransformation processes

Plants engineered to tolerate non-biological stressors such as drought, frost, high soil salinity, and nitrogen starvation were in development. In

2011, Monsanto's DroughtGard maize became the first drought-resistant GM crop to

ability to deliver bacterial EPSPS to the chloroplasts of other plants. This CP4 EPSPS gene was cloned and transfected into soybeans.

Tobacco plants have been engineered to be resistant to the herbicide bromoxynil.

Crops have been commercialized that are resistant to the herbicide glufosinate, as well. Crops engineered for resistance to multiple herbicides to allow farmers to use a mixed group of two, three, or four different chemicals are under development to combat growing herbicide resistance.

Tobacco, corn, rice and many other crops have been engineered to express genes encoding for insecticidal proteins from Bacillus thuringiensis (Bt). Papaya, potatoes, and squash have been engineered to resist viral pathogens such as cucumber mosaic virus which, despite its name, infects a wide variety of plants. The introduction of Bt crops during the period between 1996 and 2005 has been estimated to have reduced the total volume of insecticide active ingredient use in the United States by over 100 thousand tons. This represents a 19.4% reduction in insecticide use.

Many strains of corn have been developed in recent years to combat the spread of Maize dwarf mosaic virus, a costly virus that causes stunted growth which is carried in Johnson grass and spread by aphid insect vectors. These strands are commercially available although the resistance is not standard among GM corn variants.

Post-Food Crop Applications

In 2012, the FDA approved the first plant-produced pharmaceutical, a treatment for Gaucher's Disease. Tobacco plants have been modified to produce therapeutic antibodies. Algae is under development for use in biofuels. Researchers in Singapore were working on GM jatropha for biofuel production. Syngenta has USDA approval to market a maize trademarked Enogen that has been genetically modified to convert its starch to sugar for ethanol. In 2013, the Flemish Institute for Biotechnology was investigating poplar trees genetically engineered to contain less lignin to ease conversion into ethanol. Lignin is the critical limiting factor when using wood to make bio-ethanol because lignin limits the accessibility of cellulose microfibrils to depolymerization by enzymes

Companies and labs are working on plants that can be used to make bioplastics. Potatoes that produce industrially useful starches have been developed as well. Oilseed can be modified to produce fatty acids for detergents, substitute fuels and petrochemicals.

Scientists at the University of York developed a weed (*Arabidopsis thaliana*) that contains genes from bacteria that could clean TNT and RDX-explosive soil contaminants in 2011. 16 million hectares in the USA (1.5% of the total surface) are estimated to be contaminated with TNT and RDX. However *A. thaliana* was not tough enough for use on military test grounds. Modifications in 2016 included switchgrass and bentgrass. Genetically modified plants have been used for bioremediation of contaminated soils. Mercury, selenium and organic pollutants such as polychlorinated biphenyls (PCBs).

Marine environments are especially vulnerable since pollution such as oil spills are not containable. In addition to anthropogenic pollution, millions of tons of petroleum annually enter the marine environment from natural seepages. Despite its toxicity, a considerable fraction of petroleum oil entering marine systems is eliminated by the hydrocarbon-degrading activities of microbial communities. Particularly successful is a recently discovered group of specialists, the so-called hydrocarbonoclastic bacteria (HCCB) that may offer useful genes.

The number of USDA-approved field releases for testing grew from 4 in 1985 to 1,194 in 2002 and averaged around 800 per year thereafter. The number of sites per release and the number of gene constructs (ways that the gene of interest is packaged together with other elements)—have rapidly increased since 2005. Releases with agronomic properties (such as drought resistance) jumped from 1,043 in 2005 to 5,190 in 2013. As of September 2013, about 7,800 releases had been approved for corn, more than 2,200 for soybeans, more than 1,100 for cotton, and about 900 for potatoes. Releases were approved for herbicide tolerance (6,772 releases), insect resistance (4,809), product quality such as flavour or nutrition (4,896), agronomic properties like drought resistance (5,190), and virus/fungal resistance (2,616). The institutions with the most authorized field releases include Monsanto with 6,782, Pioneer/DuPont with 1,405, Syngenta with 565, and USDA's Agricultural Research Service with 370. As of September 2013 USDA had received proposals for releasing GM non-food plants, rose, tobacco, flax and chicory.

By leaving at least 30% of crop residue on the soil surface from harvest through planting, conservation tillage reduces soil erosion from wind and water, increases water retention, and reduces soil degradation as well as water and chemical runoff. In addition, conservation tillage dramatically reduces the carbon footprint of agriculture. A 2014 review covering 12 states from 1996 to 2006, found that a 1% increase in herbicide-tolerant (HT) soybean adoption leads to a 0.21% increase in conservation tillage and a 0.3% decrease in quality-adjusted herbicide use.

Chapter 10 Rates of GMO Introduction

In 2013, GM crops were planted in 27 countries; 19 were developing countries and 8 were developed countries. 2013 was the second year in which developing countries grew a majority (54%) of the total GM harvest. 18 million farmers grew GM crops; around 90% were small-holding farmers in developing countries.

Country	2013– planted (million hectares)	GM area Biotech crops
USA	70.1	Maize, Soybean, Cotton, Canola, Sugarbeet, Alfalfa, Papaya, Squash
Brazil	40.3	Soybean, Maize, Cotton
Argentina	24.4	Soybean, Maize, Cotton
India	11.0	Cotton
Canada	10.8	Canola, Maize, Soybean, Sugarbeet
Total	175.2	----

The United States Department of Agriculture (USDA) reports every year on the total area of GMO varieties planted in the United States. According to National Agricultural Statistics Service, the states published in these tables represent 81–86 percent of all corn planted area, 88–90 percent of all soybean planted area, and 81–93 percent of all upland cotton planted area (depending on the year).

Global estimates are produced by the International Service for the Acquisition of Agri-biotech Applications (ISAAA) and can be found in their annual reports, "Global Status of Commercialized Transgenic Crops".

29 countries such as the USA, Brazil, Argentina, India, Canada, China, Paraguay, Pakistan, South Africa, Uruguay, Bolivia, Australia, Philippines, Myanmar, Burkina Faso, Mexico and Spain. One of the key reasons for this widespread adoption is the perceived economic benefit the technology brings to farmers. For example, the system of planting glyphosate-resistant seed and then applying glyphosate once plants emerged provided farmers with the opportunity to dramatically increase the yield from a given plot of land, since this allowed them to plant rows closer together. Without it, farmers had to plant rows far enough apart to control post-emergent weeds with mechanical tillage.

Likewise, using Bt seeds means that farmers do not have to purchase insecticides, and then invest time, fuel, and equipment in applying them. However critics have disputed whether yields are higher and whether chemical use is less, with GM crops. See genetically modified food controversies article for information.

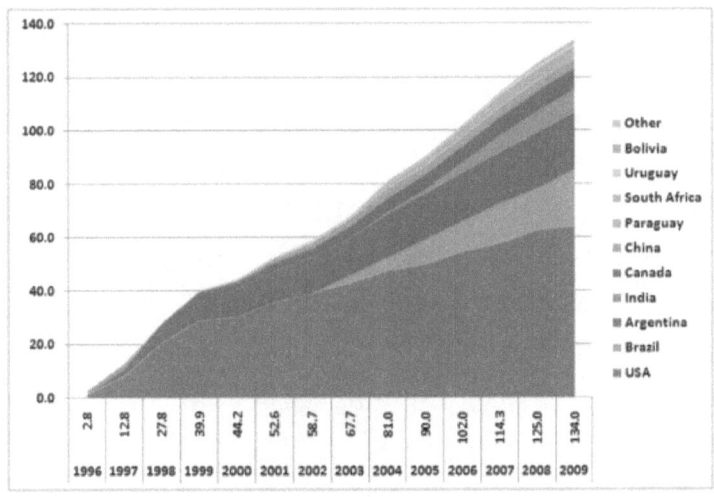

Land area used for genetically modified crops by country (1996–2009), in millions of hectares. In 2011, the land area used was 160 million hectares, or 1.6 million square kilometers

In the US, by 2014, 94% of the planted area of soybeans, 96% of cotton and 93% of corn were genetically modified varieties. Genetically modified soybeans carried herbicide-tolerant traits only, but maize and cotton carried both herbicide tolerance and insect protection traits (the latter largely Bt protein). These constitute "input-traits" that are aimed to financially benefit the producers but may have indirect environmental benefits and cost benefits to consumers. The Grocery Manufacturers of America estimated in 2003 that 70–75% of all processed foods in the U.S. contained a GM ingredient.

Europe grows relatively few genetically engineered crops with the exception of Spain, where one fifth of maize is genetically engineered, and smaller amounts in five other countries. The EU had a 'de facto' ban on the approval of new GM crops, from 1999 until 2004. GM crops are now regulated by the EU. In 2015, genetically engineered crops are banned in 38 countries worldwide, 19 of them in Europe. Developing countries grew 54 percent of genetically engineered crops in 2013.

In recent years GM crops expanded rapidly in developing countries. In 2013 approximately 18 million farmers grew 54% of worldwide GM crops in developing countries. 2013's largest increase was in Brazil (403,000 km² versus 368,000 km² in 2012). GM cotton began growing in India in 2002, reaching 110,000 km² in 2013.

According to the 2013 ISAAA brief: "...a total of 36 ';countries' (35 + EU-28) have granted regulatory approvals for biotech crops for food and/or feed use and for environmental release or planting since 1994... a total of 2,833 regulatory approvals involving 27 GM crops and 336 GM events (NB: an "event" is a specific genetic modification in a specific species) have been issued by authorities, of which:

- 1,321 are for food use (direct use or processing),
- 918 for feed use (direct use or processing) and
- 599 for environmental release or planting.

Japan has the largest number (198), followed by the U.S.A. (165, not including "stacked" events), Canada (146), Mexico (131), South Korea (103), Australia (93), New Zealand (83), European Union (71 including approvals that have expired or under renewal process), Philippines (68), Taiwan (65), Colombia (59), China (55) and South Africa (52). Maize has the largest number (130 events in 27 countries), followed by cotton (49 events in 22 countries), potato (31 events in 10 countries), canola (30 events in 12 countries) and soybean (27 events in 26 countries).

GM foods remain 'controversial' and the subject of protests, vandalism, referenda, legislation, court action and scientific disputes.continues.The controversies involve consumers, biotechnology companies, governmental regulators, non-governmental organizations and scientists. The key areas are whether GM food should be labeled, the role of government regulators, the effect of GM crops on health and the environment, the effects of pesticide use and resistance, the impact on farmers, and their roles in feeding the world and in energy production.

There is a scientific consensus that currently available food derived from GM crops poses no greater risk to human health than conventional food, but that each GM food needs to be tested on a case-by-case basis before introduction. Nonetheless, members of the public are much less likely than scientists to perceive GM foods as safe. The legal and regulatory status of GM foods varies by country, with some nations banning or restricting them, and others permitting them with widely differing degrees of regulation. No reports of ill effects have been documented in the human population from GM food. Although GMO labeling is required in many countries, the United States Food and Drug Administration does not require labeling.

Advocacy groups such as Center for Food Safety, Union of Concerned Scientists, Greenpeace and the World Wildlife Fund claim that risks related to GM food have not been adequately examined and managed, that GMOs are not sufficiently tested and should be labelled, and that regulatory authorities and scientific bodies are too closely tied to industry. Some studies have claimed that genetically modified crops can cause harm; a 2016 review that reanalyzed the data from six of these studies found that their statistical methodologies were flawed and did not demonstrate harm, and said that conclusions about GMO crop safety should be drawn from "the totality of the evidence... instead of far-fetched evidence from single studies"

Starch or amylum is a polysaccharide produced by all green plants as an energy store. Pure starch is a white, tasteless and odourless powder. It consists of two types of molecules: the linear and helical amylose and the branched amylopectin. Depending on the plant, starch generally contains 20 to 25% amylose and 75 to 80% amylopectin by **weight.** Starch can be further modified to create modified starch for specific purposes, including creation of many of the sugars in processed foods. They include:

- Maltodextrin, a lightly hydrolyzed starch product used as a bland-tasting filler and thickener.

- Various glucose syrups, also called corn syrups in the US, viscous solutions used as sweeteners and thickeners in many kinds of processed foods.
- Dextrose, commercial glucose, prepared by the complete hydrolysis of starch.
- High fructose syrup, made by treating dextrose solutions with the enzyme glucose isomerase, until a substantial fraction of the glucose has been converted to fructose.
- Sugar alcohols, such as maltitol, erythritol, sorbitol, mannitol and hydrogenated starch hydrolysate, are sweeteners made by reducing sugars.

Lecithin is a naturally occurring lipid. It can be found in egg yolks and oil-producing plants. It is an emulsifier and thus is used in many foods. Corn, soy and safflower oil are sources of lecithin, though the majority of lecithin commercially available is derived from soy.] Sufficiently processed lecithin is often undetectable with standard testing practices.] According to the FDA, no evidence shows or suggests hazard to the public when lecithin is used at common levels. Lecithin added to foods amounts to only 2 to 10 percent of the 1 to 5 g of phosphoglycerides consumed daily on average. Nonetheless, consumer concerns about GM food extend to such products. This concern led to policy and regulatory changes in Europe in 2000, when Regulation (EC) 50/2000 was passed which required labelling of food containing additives derived from GMOs, including lecithin[1] Because of the difficulty of detecting the origin of derivatives like lecithin with current testing practices, European regulations require those who wish to sell lecithin in Europe to employ a comprehensive system of Identity preservation (IP).

The US imports 10% of its sugar, while the remaining 90% is extracted from sugar beet and sugarcane. After deregulation in 2005, glyphosate-resistant sugar beet was extensively adopted in the United States. 95% of beet acres in the US were planted with glyphosate-resistant seed in 2011. GM sugar beets are approved for cultivation in the US, Canada and Japan; the vast majority are grown in the US. GM beets are approved for import and consumption in Australia, Canada, Colombia, EU, Japan, Korea, Mexico, New Zealand, Philippines, Russian Federation and Singapore. Pulp from the refining process is used as animal feed. The sugar produced from GM sugarbeets contains no DNA or protein—it is just sucrose that is chemically indistinguishable from sugar produced from non-GM sugarbeets. Independent analyses conducted by internationally recognized laboratories found that sugar from Roundup Ready sugar beets is identical to the sugar from comparably grown conventional (non-Roundup Ready) sugar beets. And, like all sugar, sugar from Roundup Ready sugar beets contains no genetic material Livestock and poultry are raised on animal feed, much of which is composed of the leftovers from processing crops, including GM crops. For example, approximately 43% of a canola seed is oil. What remains after oil extraction is a meal that becomes an ingredient in animal feed and contains canola protein. Likewise, the bulk of the soybean crop is grown for oil and meal. The high-protein defatted and toasted soy meal becomes livestock feed and dog food. 98% of the US soybean crop goes for livestock feed. In 2011, 49% of the US maize harvest was used for livestock feed (including the percentage of waste from distillers grains). "Despite methods that are becoming more and more sensitive, tests have not yet been able to establish a difference in the meat, milk, or eggs of animals depending on the type of feed they are fed. It is impossible to tell if an animal was fed GM soy just by looking at the resulting meat, dairy, or egg products. The only way to verify the presence of GMOs in animal feed is to analyze the origin of the feed itself."

A 2012 literature review of studies evaluating the effect of GM feed on the health of animals did not find evidence that animals were adversely affected, although small biological differences were occasionally found. The studies included in the review ranged from 90 days to two years, with several of the longer studies considering reproductive and intergenerational effects.

The use of genetically modified food-grade organisms as recombinant vaccine expression hosts and delivery vehicles can open new avenues for vaccinology. Considering that oral immunization is a beneficial approach in terms of costs, patient comfort, and protection of mucosal tissues, the use of food-grade organisms can lead to highly advantageous vaccines in terms of costs, easy administration, and safety. The organisms currently used for this purpose are bacteria (Lactobacillus and Bacillus), yeasts, algae, plants, and insect species. Several such organisms are under clinical evaluation, and the current adoption of this technology by the industry indicates a potential to benefit global healthcare systems.

Concluding Remarks

We have remarked on four Green Revolutions, and each continues through the time of this writing. The fact is that we can manage, through Abundance Theory application, to improve the nutritional and cultural conditions across the planet. It will take political will, technological advancement and public education to achieve more than we are able to provide today.

The author invites an exchange with any reader.

About the Author

Mark Roberts-Seymour, B.A.Sc., P.Eng., CD, ACG, OFS is a Canadian Professional Forensic Engineer, a recognised stoneworks conservator, Chartered Demographer, non-fiction author, lay-brother, professional public speaker (Toastmaster Advanced Communicator Gold), technical paper referee, statistical assessor, and editor.

Over succeeding years his publications have included:

- *Old, Unemployed and Pissed:Late Career Canadians Coping with Long-term Unemployment*
- *Restructuring a Broken Canadian Economic-Democracy*

- *Three Proto-Christian Orthodoxies, The Gospel of Paul, Alexandrian Orthodoxy and Proto-Christian Gnosticism: A Comparison*
- *Clinical Happiness: Measurement, Measures, Goals and Habit Conditioning*
- *Anglicanism: From Henry to Henrietta*
- *Cyborg:* Smartphone Reliance, AI and Transhumanism
- *An Afterlife: Who Cares! Quotations on an After-life with Biographical notes*
- *Conservation of Heritage Cemeteries,*
- *Green Revolutions – Will they be Enough,*
- *Suburban and Ex-Urban Self Sufficiency,*
- *Life Extension for Seniors Book 1: Expectations, Ageing Inhibition, Nutrition, Brain 'Wiring', Habits, Exercise and Ethics*
- *Old, Unemployed and Pissed: Late-Career Canadians Coping with Long-Term Unemployment*
- *The Working Poor: Who are Poor and What Can Be Done: A Central British Columbia Case Study*

- *Life Extension for Seniors Book 2: Methods, Research, Supplementation, Health and Life-Prolonging Strategies*
- *Abrahamic Gnosticism is not Scary,*
- *Proofs of God – Philosophy of Religion and Science Converge,*
- *Anglicanism – From Henry to Henrietta,*
- *Happiness: Biochemistry, Goals and Habituation,*
- *The Parable of the Prodigal Son - Death, Rebirth, Recognition and Reconciliation,*
- *Nutrition for Older Workers,*
- *Nutrient Supplements for the Older Worker,*
- *Canadian Systems: Changing our Economic-Democracy,*
- *Gnosticism as Revelation: from St. Paul to C.G. Jung,*
- *Sabotage, Wealth and New Classes,*
- *Christian Metempsychosis: Elijah and John the Baptist,*
- *Sanctification – It's for Everyone!,*
- *The Gospel of Paul, Christian Gnosticism and Alexandrian Orthodoxy – A Comparison,*

Green Revolutions

- *An Afterlife, Who Cares!* **Quotations from 231 sources**
- *Renovating the Canadian Economic-Democratic System,*
- *To Coin a Phrase or Not to Coin a Phrase: Cliches, Metaphors and Euphemisms in Use,*
- *Canadian Systems: Changing our Economic-Democracy,*
- *It Can't Happen Here: Future Mechanisation, Despair and Suicide*

To access further information on these books [or for purchase], links to the Distributor Ammazon.com are indicated in the list preceding.

Mark acted for more thirty years as a Registered Professional Engineer (P.Eng., PE, ing.), technical author and editor for: The Government of BC (Lands Forests and Water Resources), BH Levelton and Associates, Warnock Hersey Professional Services, Heritage Technologies Press, RM Hardy and Associates, G.W. Spratt Limited and others. Mark also owned and managed a private Materials Engineering firm for an additional ten years (Roberts Seymour and Associates Limited).

He remains active in several service, political and social justice organisations as well as maintaining his professional standings. He is married, the father of four adult children and many grandchildren, and calls Vernon, British Columbia, Canada his home. Mark Roberts-Seymour can be reached directly by email at merscanada@gmail.com and by telephone at (250) 306-0550. For keynote speaking and seminar leadership contact (250) 721-5683.

www.ingramcontent.com/pod-product-compliance
Lightning Source LLC
Chambersburg PA
CBHW030644220526
45463CB00004B/1636